CELEBRATING

50 YEARS

Texas A&M University Press
publishing since 1974

LESSONS FROM
LEOPOLD

Myrna and David K. Langford Books on Working Lands

LESSONS FROM LEOPOLD

LEARNING FROM THE LAND

Steve Nelle

with Iliana Peña

Color photography by Wyman Meinzer

FOREWORD BY DAVID K. LANGFORD
FOREWORD BY ANDREW SANSOM

TEXAS A&M UNIVERSITY PRESS | College Station

∞ This paper meets the requiremefnts of ANSI/NISO Z39.48–1992
(Permanence of Paper).
Binding materials have been chosen for durability.

Manufactured in China through Martin Book Manufacturing.

Library of Congress Cataloging-in-Publication Data

Names: Nelle, Steve (Stephan A.), author. | Peña, Iliana, author. |
Meinzer, Wyman, photographer (expression) | Langford, David K., 1942–
writer of foreword. | Sansom, Andrew, writer of foreword.
Title: Lessons from Leopold: learning from the land / Steve Nelle with
Iliana Peña; color photography by Wyman Meinzer; foreword by David K.
Langford; foreword by Andrew Sansom.
Other titles: Myrna and David K. Langford books on working lands.
Description: First edition. | College Station: Texas A&M University Press,
[2024] | Series: Myrna and David K. Langford books on working lands |
Includes bibliographical references and index.
Identifiers: LCCN 2024029453 (print) | LCCN 2024029454 (ebook) | ISBN
9781648432484 (hardcover) | ISBN 9781648432491 (ebook)
Subjects: LCSH: Leopold, Aldo, 1886-1948—Influence. | Conservation of
natural resources. | Wildlife management. | Environmental ethics. | Land
use—Environmental aspects. | BISAC: NATURE / Environmental Conservation
& Protection | NATURE / Essays
Classification: LCC QH75 .N385 2024 (print) | LCC QH75 (ebook) | DDC
333.95/16—dc23/eng/20240701
LC record available at https://lccn.loc.gov/2024029453
LC ebook record available at https://lccn.loc.gov/2024029454

To my lovely and devoted wife,
Marnie,
who understands me like no one else
and loves me anyway

CONTENTS

Chapter 6. Lessons in Land Management 115

Chapter 7. Lessons in Wildlife Management 144

FOREWORD

One of Aldo Leopold's most-quoted land management observations centers on tools. In 1933, at the age of forty-six, in his book *Game Management*, Leopold wrote, "The central thesis of game management is this: game can be restored by the creative use of the same tools which have heretofore destroyed it—axe, plow, cow, fire, and gun."

In addition to the obvious wisdom and practicality of the statement, I am struck by two other things. First, Leopold did not burst upon the scene as a precocious teenage prodigy. Instead, he grew into a conservation pioneer, who eventually earned the moniker the "Father of Wildlife Management," by feeding his insatiable curiosity about the natural world through observation, experience, and lifelong learning.

Second, he perhaps left out a crucial tool—the pen. For instance, the pen once popularized ornamental plumage on women's hats, which drove legions of shorebirds to the brink of extinction. Years later, the pen swayed public opinion and made such plumage anathema, the mark of the ill-informed and uncaring. As a result, mass slaughter stopped and populations rebounded.

But even more to the point of Leopold's life, imagine the North American landscape and the state of private stewardship if Leopold had kept his musings to himself. Maybe someone else would have moved to the forefront of the public conversation and spurred a land ethic, but maybe not.

Fortunately for us all, Leopold chose to nurture his own land, apply his intellect to big questions, and put his pen to paper. The potent combination created a legacy that shaped not only our nation's land but also our collective conservation philosophy.

As a former director of the Sand County Foundation, which is a nonprofit organization based in Wisconsin dedicated to keeping Leopold's work vibrant and alive, I had a bird's-eye view of his impact sixty years after his death. In 2008 we were meeting in Madison just after the second Noah's flood occurred. The foundation's chairman of the board and executive director hired a plane to fly us over the area near "the Shack," the homestead on Leopold's farm of

Aldo Leopold and his bride Estella, anticipating their future together, unaware of the lifetime of experiences and reflections that would change the conservation landscape of America forever. *Courtesy of the Aldo Leopold Foundation and University of Wisconsin–Madison Archives*

eighty acres when originally purchased. From just north of the Shack, the Wisconsin River's water was a muddy soup except for the half-mile where it ran past Leopold's farm. That small stretch was clear because Leopold practiced what he preached. Apparently, the majority of his neighbors up and down the river didn't follow the same practices.

I think this, among many things, is what made Leopold's words and work speak to so many conservationists like Steve Nelle, who followed in Leopold's footsteps and managed to broaden his trailblazing path with their own work. I've been fortunate to claim Steve as a friend and colleague for more years than either of us cares to count. From the earliest days of my tenure as CEO of the Texas Wildlife Association, Steve served as a touchstone for me. As a thirty-five-year veteran of the Natural Resources Conservation Service who worked directly with landowners across the state, Steve often recognized emerging issues before anyone else got the news.

Our paths first crossed on field days when he educated landowners on myriad topics. After hearing him speak, I always made it a point to seek out his presentations because I knew that, like Leopold, his observations and strategies were grounded in experience. Although Steve doesn't pound a bully pulpit or use soaring rhetoric to play the audience, his passion is evident in his quiet commitment to the land, the wildlife, and the people who manage them. He inspires by knowledge and example. I suspect that he, as a lifelong learner, has forgotten more about plants, soil, water, livestock, wildlife, and their interrelationships than all but a very few have ever known.

Fortunately for us all, Steve carries a pen in his toolbox. When I read the first "Lessons from Leopold" essay in the *Texas Wildlife* magazine about ten years ago, I suspected it was the start of something good. After I read the first three or four installments, I knew it was something exceptional. I immediately picked up the phone and called Steve to ask when he was going to write a book. He replied immediately, "Not now."

Through the years, we repeated the conversation periodically. Sometimes on the phone and other times via email. I asked the same question, and the response was always the same: an unequivocal "Not now."

Fast-forward to September 2021, when I read "Uneasy Conservation," Steve's then-latest installment in the "Lessons from Leopold" series about the need to balance the conservation of our natural resources with the needs of our expanding economy. I asked him again about writing a book, fully expecting to be rebuffed. Much to my delighted surprise, I almost immediately got an email reply saying "yes." I got the ball rolling with the Texas A&M University Press before he could change his mind.

I don't think it's hyperbole to offer this book, *Lessons from Leopold*, as the Lone Star companion for the venerated works of Leopold. Steve learned the lessons from Leopold well, applied them to the far reaches of Texas, and left the state of land stewardship better for it.

David K. Langford
July 25, 2023
Laurels Ranch, Comfort, Texas

FOREWORD

Among my most memorable days since moving to the Hershey Ranch in the Texas Hill Country are those I've spent walking across the countryside with Steve Nelle. He is the colleague I consider to be the most respected range scientist in Texas and is not only able to identify every plant on the landscape, but he will often enthrall me and anyone else nearby with a description of a particular plant's interaction with other grasses or flowers and how this interaction allows both to flourish despite browsing by deer or grazing by cattle.

Here, with help from Iliana Peña, Steve engages us with many remarkable observations from another legend and hero of the conservation movement in America, Aldo Leopold. These essays have been produced over the years by Nelle as a regular and key component of *Texas Wildlife*, the magazine of the Texas Wildlife Association. The association is among the largest and most influential conservation organizations in the state, and Steve's presence in its publication has both strengthened and celebrated its commitment to land stewardship.

The leading event honoring Texas landowners for their stewardship is the Texas Land Steward Awards, presented by the Texas Parks and Wildlife Department. In 2016, Steve Nelle was presented with Special Recognition–Landowner Outreach and Education at this prestigious awards event for his contributions to enlightened land stewardship across the state. As we explore and enjoy *Lessons from Leopold*, it is easy to understand why. It is an honor for it to be part of the Texas A&M University Press series on working lands.

Andrew Sansom

PREFACE

This book is a collection of fifty-five short lessons inspired by the writings of Aldo Leopold. Each lesson has two parts: a quotation or passage from Leopold followed by commentary on how it relates to modern-day land management.

These lessons are a way of dividing the voluminous writings of Leopold into smaller, more digestible portions. Somewhat like the proverbial eating of an elephant, it can only be done one bite at a time.

The lessons merely sweep the surface of Leopold's writings. Like nuggets of gold lying on top of the ground, most of these lessons are obvious and easy to pick up. With each lesson, there is more that could be said, but the intent is to be brief and concise. Despite their brevity, the lessons in whole give a solid representation of Leopold and his philosophies about land management, especially the role of private lands stewardship.

The interested reader may want to dig deeper to discover other facets of Leopold and their application to contemporary land management. If these lessons whet the appetite for more, they will serve a good purpose. But even if the reader stops here and goes no further, these lessons will stimulate deeper thinking of our relationship to the land and how we can get along with nature better.

The excerpts for each lesson were chosen because I have personally observed their truth and relevance for today's private landowners and anyone interested in conservation. These are not just theoretical truths. I have seen these principles at work on the many private ranches I have worked with since 1976.

These are practical lessons that if properly applied will help private landowners and land managers better understand and better care for the land under their trust. This is the essence of land stewardship. When land is taken care of, it provides greater benefits and many beneficiaries. These benefits, tangible and intangible, accrue to the landowner, their neighbors, future generations, and all of society. Everyone benefits when land is well taken care of. While the actual term "land stewardship" is not found in Leopold's writings, the concept of stewardship is woven throughout his essays and is perhaps the overarching theme of his life. Fostering this stewardship is a key purpose of the book.

The lessons presented here can be read and understood individually, but the collection is best considered in whole. The lessons reinforce each other and have a synergistic, compounding effect when linked to other lessons. Readers will immediately notice that there is considerable overlap and reiteration between lessons and chapters, and that is by design. This is in keeping with the nature of ecology and conservation, where everything is interconnected and nothing stands alone. When concepts and truths are repeated or reinforced it emphasizes their importance.

Everyone knows the joys and the sorrows of one's relationships with other people. Many people also understand the relationships that people can have with animals, whether pets, livestock, or wild animals. What is less well understood is the relationship that exists between people and the land. Like other relationships, it can be healthy or unhealthy. By understanding what Leopold taught, our relationship to the land will grow stronger, healthier, and more mutually beneficial.

It is not the goal of this book to put Leopold on a pedestal. He made mistakes just like the rest of us. He sometimes spoke prematurely and expressed his opinions as though they were proven facts. He did not live long enough to have some of his mistakes refuted while he was still alive, but looking back we can see some flaws.

Despite his imperfections, Leopold was outstanding in his ability to see the big picture of the land, which he defined as the sum total of soil, water, plants, animals, and people as well as all of the interactions among and between these. Leopold was insistent that people are an integral part of the land, not bystanders. He was able to write in a way that made people think more clearly about their attitudes and actions with regard to the land.

In Leopold's day, land was often not revered by landowners. It was generally regarded as a way to make a living or a place to hunt, not as something to love and respect. When mismanagement took place, it was considered the inevitable cost of working the land and producing commodities. Land abuse was commonplace, and the new discipline of conservation was in its infancy.

In the years since his writing, the practice of conservation and the ethics of private land stewardship have grown. No doubt Leopold with the help of others has influenced today's landowners and land managers either directly or indirectly. It is my hope that these lessons might broaden or deepen the growth of stewardship even further.

The intended audience for this book is broad. I hope that anyone who appreciates nature, natural resources, and how they are managed will find these lessons interesting and inspiring. Those who actually own or manage land will find more direct and deeper relevance, but even for others these lessons will provide plenty of food for thought regarding the miracles of nature and our relationship to the land.

These lessons are written from my perspective of conservation and land management in Texas where I have lived and worked. Although the book is Texas-centric in context, the lessons, like Leopold's writings, apply to a much wider setting. In fact, many lessons are universal in applicability.

Others who read and study Leopold, who live in different places with different backgrounds, will certainly come away with some different lessons and different applications. So if this book serves as a stepping-stone for your own journey with land and Leopold it will be successful.

The photos in this book are a vital part of the whole message. The historic black-and-white images of Leopold in chapter 1 allow us to see him in his element. The color images by Wyman Meinzer in chapters 2 through 7 are outstanding in their ability to convey the land stewardship message. They are more than pretty pictures to grab attention. In fact, some are not pretty at all. The photos help tell the story of the land and humankind's relationship with the land.

My hope is that you will use this book, not merely read it. Mark the pages and make notations. Note where you see things differently. Add your own observations and examples. If this book ends up being worn, tattered, dog-eared, and coffee-stained, it will give evidence of having been useful.

ACKNOWLEDGMENTS

This book is the result of many people having actively helped me throughout my entire life. I am profoundly indebted to family, friends, landowners, coworkers, and mentors who have invested in my life and my life's work of conservation.

My parents, George and Loraine Nelle, have supported, helped, and encouraged me ever since I came into this world. They provided many opportunities that fostered my love of nature and how to take care of it. Mom conveyed the appreciation and joys of nature derived by walking slowly and watching carefully, with the curiosity to notice even the small, simple things. Dad instilled the importance of hard work, perseverance, and always giving your best in everything you do, no matter what. They have molded my life in ways beyond measure.

My brother, Doug Nelle, was my hunting and fishing buddy growing up, and that was a major part of our boyhood. We share many fond memories of time together on the river or afield with shotguns, and I now admit that Doug was always a better hunter and fisher than I. He is one of my heroes. My sisters, Laurel and Caryl, do not hunt or fish, but they appreciate nature in other ways and have been kind and gracious to me when I did not deserve it.

Each of my grandparents contributed to my love of the land in unique ways. Grandpa (Walter Herman Victor Stephan) taught Doug and me how to fish and took us fishing countless times. He also had a massive garden where we learned what it means to work the land. Grandma (Valesca Wollshlaeger Stephan) was a sweet and gentle person who loved birds and flowers. But she scolded us if we ever pointed our BB guns at anything but a tin can. Granddaddy (George Anderson Nelle) picked cotton as a boy and raised rabbits, chickens, and pecans later in life. He was always cultivating and tinkering, always jovial and fun-loving. Grandmother (Lucille Conn Nelle) loved birds and flowers and grew plants of all kinds. She nurtured them just like she nurtured her grandchildren. My grandparents are all enormously special to me, and I like to think that pieces of them live on in me.

My wife, Marnie Needham Nelle, deserves special recognition for tolerating wild animals in the freezer and bones, skulls, antlers, feathers, fur, owl pellets, snakeskins, dried grasses, seeds, rocks, fossils, and other natural artifacts all over the house. She has put up with my eccentric ways and never stopped loving me. She is my best friend.

Our three children, Will, Andrea and Stephan all enjoy nature in various ways, and each has lived in the country and cared for their own piece of land, giving me the satisfaction that they value their connections to the land. They are fine parents to our eight beloved grandchildren. Every father wants his children to surpass him in character and right living, and this desire has come to pass for me.

I have been fortunate to work with many of the finest natural resource professionals in Texas who have helped shape my understanding of the land and how to manage and conserve it. They have given me an education that far surpasses any formal schooling. Al Brothers and Murphy Ray each took me under their wing when I was a young conservationist and taught me about the proper integration of wildlife management and cattle ranching. Val Lehmann (who personally worked with Leopold) gave me encouragement and direction early in my career. Dan Caudle and Gary Valentine have been primary mentors and provided much-needed guidance and motivation over several decades; they are the epitome of professionalism and integrity. John L. "Chip" Merrill inspired and motivated me beyond words and embodied character and genuine land stewardship. Russ Pettit, Donnie Harmel, Kent Mills, Chris Farley, Stan Reinke, Dale Rollins, Butch Taylor, Darrell Ueckert, Greg Simons, Ricky Linex, Wayne Gabriel, Gene Miller, Kent Ferguson, Mark Moseley, Elizabeth McGreevy, Hugh Aljoe, Allan Savory, and Bill Eikenhorst have all sharpened me and challenged me in countless ways.

I have flown thousands of hours of low-level wildlife surveys on millions of acres with Jeff Hill and the late Mackey McEntire of Concho Aviation. They have been rich sources of practical information as well as superb pilots and personal friends. Wayne Elmore, Kenneth Mayben, Sky Lewey, and Janice Staats have taught me about creeks, rivers, and riparian areas and how we can better understand and conserve these special parts of the landscape.

I am obliged to and appreciate hundreds of private landowners and ranch managers across Texas for teaching me what real conservation and stewardship are. They have invited me to their ranches for conservation assistance, but in many cases I am the one who gained more benefit. Bert Gallagher, Gene S. Walker Sr., and Frank Matthews were especially helpful to me in the early years in South Texas. Phillip Robbins, Charley Christensen, Frank Price, Skipper Duncan, Rob Hailey, Kenneth Barbutti, Will Harte, Robert Potts, and Ryland and Anson Howard have taught me much about practical range and wildlife management and land stewardship in West Texas. Hal and Amy Zesch, Charlie Granstaff, Ward Whitworth, Chad Druen-Miller, Andy Sansom, Kevin Wessels, Kim and Pam Bergman, and Richard and Josephine Smith in the Hill Country region have each imparted to me valuable lessons in caring for the land. These and many other landowners have truly been my teachers, and I have tried to soak up the lessons they taught. By listening to them and observing their management I have gained insights into the practical conservation of soil, water, plants, and animals. My role as conservationist has largely been to pass along what I have learned from others.

The many landowners in the Riley Mountain and Prairie Mountain Wildlife Management Associations in Llano County, Texas, have helped me see both the values and challenges of managing shared wildlife resources at the landscape level.

My interest in Leopold may never have developed were it not for Thelma Wilson, my high school English teacher, who assigned a book report in 1971. She sensed my interest and connection with the natural world upon reading *A Sand County Almanac* and gently encouraged me to continue with what she perceived to be a calling.

David K. Langford inspired the idea for this book and has nudged me along patiently each step of the way. Langford has been the prime ambassador of the private lands stewardship message for many years, and his message has rubbed off hard on me. I very much appreciate his friendship and mentorship.

The lessons contained here were previously published in *Texas Wildlife* magazine by the Texas Wildlife Association (TWA), and I am indebted to the TWA for providing that platform to communicate the timeless land management philosophies and principles of Aldo Leopold. The TWA is, without exaggeration, the most successful and effective conservation organization in Texas, and it is an honor to be associated with them.

I am indebted to Iliana Peña, Peña Conservation Strategies, now working with Knobloch Family Foundation, for writing the introduction and for her help from the very start in guiding the development of this book; her input has been instrumental and much appreciated.

The Aldo Leopold Foundation and University of Wisconsin–Madison Archives granted permission to use the historic photos in chapter 1. Wyman Meinzer, whom I have known and respected since 1971 during our days at Texas Tech, provided the remarkable color photos that grace the book.

Marguerite Avery, Katie Duelm, Shannon Davies, Lorie Woodward, Kim Rothe, Burt Rutherford, and Lee Young have all helped in various ways to make this book better than it would have been without them. Chris Dodge skillfully and patiently copyedited this book, adding value and clarity to each page. Curt Meine has helped sharpen my understanding of Leopold and the background of some of his essays.

As a follower of Jesus Christ, my life as a Christian is closely tied to my profession. I acknowledge what the Bible proclaims, that everything we call nature was created by God and that He delegated the responsibility of stewardship to humans, to care for and derive sustenance from it. Jim Haesemeyer and Bob Hall have been primary spiritual mentors, teachers, and examples for many years, and I am indebted to them for demonstrating the practical and eternal value of trusting God in all things. It is only by God's enabling grace that I have been able to succeed in my profession and to be gaining some feeble understanding of the workings of nature.

INTRODUCTION

BY ILIANA PEÑA

In Texas, where the natural spirit of the land thrives alongside a rich assemblage of wildlife and agriculture, one person's experiences and writings demonstrate the lasting legacy of Aldo Leopold's conservation ethos. Welcome to *Lessons from Leopold* by Steve Nelle. This compilation of thought-provoking lessons weaves together timeless passages and quotes from Leopold's writings with Nelle's experienced personal narratives, scientific knowledge, and practical insights to shed light on the unique opportunities and challenges of land stewardship in the Lone Star State. The writings began as a bimonthly column for *Texas Wildlife*, the monthly magazine of the Texas Wildlife Association. For TWA members, the column quickly became an anticipated must-read. With encouragement from respected colleagues, landowners, and friends, Nelle brings these lessons together formally to make them accessible to a much larger and broader audience, including everyone from the nature-loving small landowner at the edge of town to the largest of agricultural producers. Equally, this book is for the hunter, fisher, birder, student, educator, natural resource professional, Master Naturalist, nature lover, environmental advocate, and so on. This book is for anyone who feels they would benefit from a better understanding of Leopold's ideas and humanity's relationship to the land.

Steve Nelle, a revered voice in land management and conservation, has dedicated his life to observing, studying, and advocating for the delicate balance between the needs of people and the workings of nature. Through his engaging and enlightening articles, Nelle has captivated readers with his ability to draw on Aldo Leopold's teachings and apply them to Texas's unique landscapes and challenges. As a trusted voice in the field, Nelle shares his firsthand experiences, lessons learned, and visionary perspectives, enriching our understanding of Leopold's land ethic in the context of Texas's diverse ecosystems.

Nelle unveils in these pages the interconnectedness between wildlife, habitat, and responsible land management. Through his poignant storytelling, we witness the sometimes awkward dance between progress and preservation, the challenges faced by landowners, and the innovative solutions that arise from embracing a rounded approach to conservation.

Drawing inspiration from Leopold's groundbreaking *A Sand County Almanac* and other writings, Nelle challenges us to view the land not as a simple commodity but as a living, breathing entity that demands our care, respect, and understanding. He eloquently asks us to consider the impacts of our actions, highlighting the importance of land management practices that prioritize ecological health and the well-being of both wildlife and people.

Lessons from Leopold serves as a guidebook offering practical advice, thought-provoking reflections, and actionable steps toward sustainable land stewardship. Through these lessons, we gain a deeper appreciation for the necessary but challenging balance between human needs and the intricate web of life, encouraging us to embrace a sense of responsibility and reverence for the natural world.

As we embark on this informative and enlightening journey, we recognize the powerful ripple effect that Leopold's teachings can have on shaping the consciousness of conservation across Texas. Through Nelle's invaluable perspectives and experiences, we are inspired to continue the legacy of Aldo Leopold, forging a future in which people, wildlife, and the land thrive and coexist harmoniously. The insights and practical wisdom in this book are a testament to the enduring influence of Leopold's land ethic and the dedicated efforts of individuals like Steve Nelle in conserving the natural wonders of the Lone Star State.

LESSONS FROM
LEOPOLD

Leopold resting on one of his hunting trips to the Gavilan River region of Mexico in 1938. Leopold the philosopher-hunter-ecologist frequently took the time to think and contemplate without distraction. *Courtesy of the Aldo Leopold Foundation and University of Wisconsin–Madison Archives*

1

WHO IS

ALDO LEOPOLD?

To ask "who was Aldo Leopold?" might be more correct than asking "who is Aldo Leopold?" But Leopold is still actively shaping our thoughts about the land even though he died in 1948. Thus, he remains alive in this book, with the clarity and prophetic relevance of the written record he left behind.

Perhaps no one has left a more lasting and significant mark on the science, art, and ethics of land management as Leopold. Although he is highly esteemed in certain land management circles, he is less well-known in others. The following short sketch of his life is an introduction for those who don't know him.

Leopold was a hybrid of naturalist, hunter, fisherman, conservationist, ecologist, landowner, father and husband, educator, student, philosopher, writer, and leader. The hybridization of these created a rare and remarkable person.

As a naturalist, he grew to understand soils, botany, ornithology, forestry, rangelands, wildlife habitat, and outdoor skills in general. He was very much at home in the outdoors whether it was with shotgun, fly rod, binoculars, notepad, or none of these. He extracted great value from his relationship with the land, and he gave back even greater value to those who would someday read his works.

Leopold was born in 1887 in Burlington, Iowa, to Carl and Clara Leopold as the eldest of four children in a home overlooking the Mississippi River. He grew up exploring with his younger brother Carl Jr. and hunting and fishing with his father while developing a deep attachment to nature and all its elements. His father instilled in his sons the principles of sportsmanship, woodsmanship, and the ethics of restraint, and Leopold honed these his entire life.

Leopold examines his nursery of young tamarack trees grown from seed on his farm near Baraboo, Wisconsin. Leopold's formal education was in forestry, and he always had a deep interest in growing trees. *Courtesy of the Aldo Leopold Foundation and University of Wisconsin–Madison Archives*

Leopold with wife Estella Bergere Leopold, circa 1930s. She was the epitome of a devoted wife and mother as well as an occasional hunting and fishing partner. *Courtesy of the Aldo Leopold Foundation and University of Wisconsin–Madison Archives*

His father owned the Leopold Desk Company, building quality hardwood desks from oak, walnut, and cherry. Perhaps it was natural for Aldo to pursue an education in forestry. He enrolled in Yale in 1906 and graduated with a master's degree in forestry in 1909.

Fresh out of Yale, young Leopold went to work for the US Forest Service on public lands in the Southwest. While in New Mexico he met his future wife Estella Bergere, of Santa Fe, and they were married in 1912. Estella was cultured, but she was also an avid outdoorswoman and an accomplished archer. They would have five children: Starker (born in 1913), Luna (1915), Nina (1917), Carl (1919), and Estella (1927). Family was obviously important to Leopold, and he wrote a great deal about them and enjoyed many hunting, fishing, and camping trips with them. Each of Leopold's children went on to contribute professionally to the world of natural resource conservation in significant ways.

Leopold with middle child Nina in 1940 at the farm. Leopold was actively and deeply involved in the lives of all of his five children. *Courtesy of the Aldo Leopold Foundation and University of Wisconsin–Madison Archives*

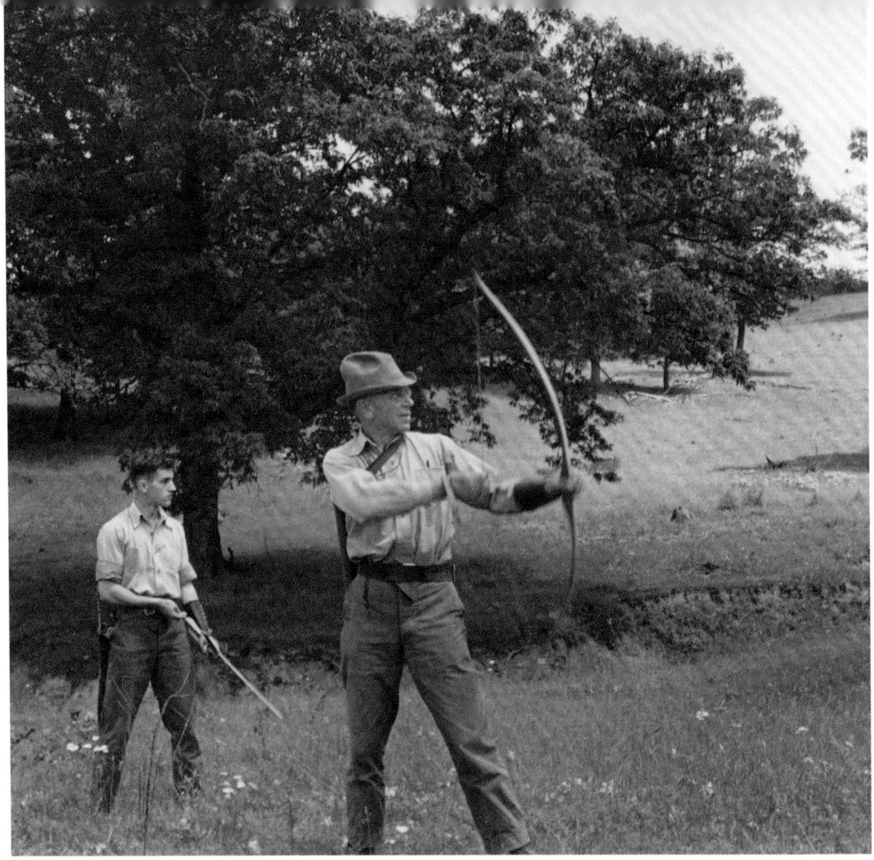

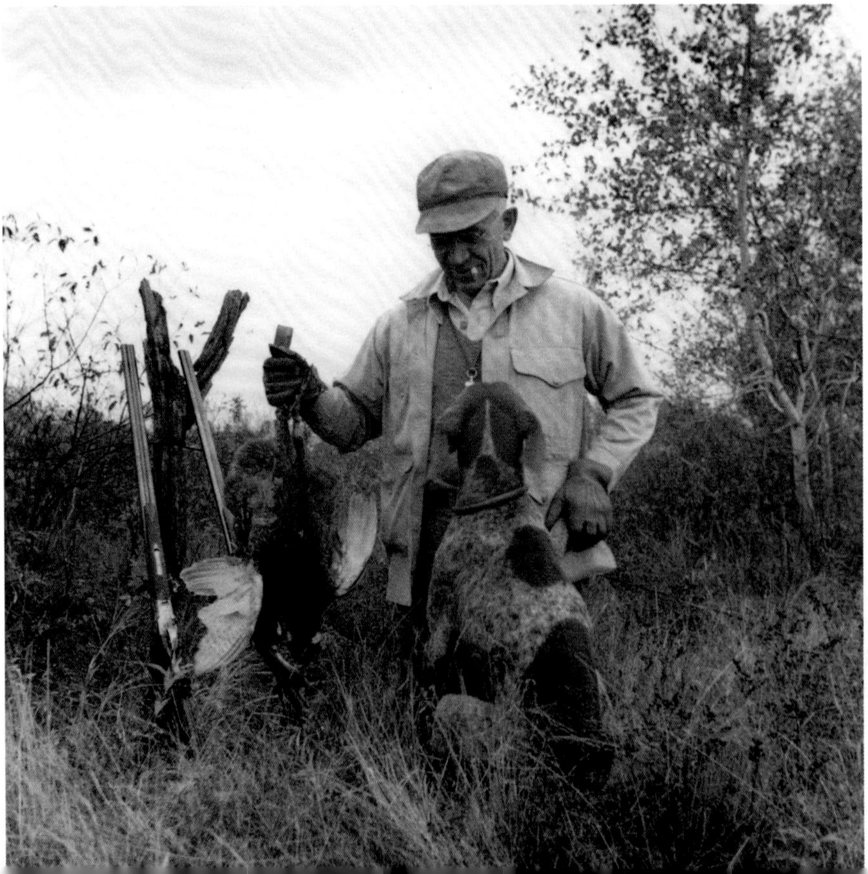

Facing top, Leopold and son Carl practicing their archery skills in about 1935. Leopold was an avid bow hunter. *Courtesy of the Aldo Leopold Foundation and University of Wisconsin–Madison Archives*

Facing bottom, Leopold with his bird dog Gus in 1943. Hunting was Leopold's primary connection to the land. He enjoyed it to the fullest for most of his life. *Courtesy of the Aldo Leopold Foundation and University of Wisconsin–Madison Archives*

Above, Leopold holding a lake trout in the Boundary Waters in 1924. Leopold and his family enjoyed fishing, especially fly fishing. He understood the relationship between the management of land and the health of the waters. *Courtesy of the Aldo Leopold Foundation and University of Wisconsin–Madison Archives*

Young Leopold at his first assignment in Apache National Forest in Arizona in 1911. Leopold became skilled in the new science of range management and understood the relationship between grazing and the condition of watersheds and wildlife habitat. *Courtesy of the Aldo Leopold Foundation and University of Wisconsin–Madison Archives*

To say that Leopold was an avid hunter and fisher is an understatement. Hunting and fishing is what defined young Leopold, and the love continued his entire life. Without his hunting and fishing, Leopold would not exist as we know him. He certainly cultivated other interests and skills, but the Leopold we know was forged in large part by hunting in fall and winter and fishing in spring and summer. It was his primary connection to the land, especially in the early years.

Leopold worked for the Forest Service from 1909 to 1928 in Arizona, New Mexico, and Madison, Wisconsin (the last for four years), and then he worked in private wildlife management consulting for five years. In 1933 he was selected to become the nation's first professor of wildlife management, at the University of Wisconsin, and later he became chairman of the school's Department of Wildlife Management, then newly formed, a position he would hold until his death in 1948.

Leopold (*second from left*) teaching his students the safe and effective use of fire as a habitat management tool. Leopold was the nation's first professor of wildlife management. *Courtesy of the Aldo Leopold Foundation and University of Wisconsin–Madison Archives*

Leopold's various jobs were never his passion but rather ways to direct and use his passion. Leopold's passion was the land. It did not matter if it was a brushy gravel bar on the Rio Grande hunting Gambel's quail, a weeklong canoe trip with his sons, or working on his farm. Leopold's avocation was also his vocation, a good lesson for all of us for a successful career.

For the first twenty years of his career, Leopold worked almost exclusively on federally owned public lands. While he enjoyed his work on Forest Service land and accomplished much, he seemed to find greater fulfillment and purpose in his later career working mostly on private lands and with private

Leopold with his dog Gus at the Riley Game Cooperative, where a group of adjacent landowners worked together to manage habitat. In the latter half of his career, Leopold worked almost exclusively with private landowners and learned the joys and challenges of private lands stewardship. *Courtesy of the Aldo Leopold Foundation and University of Wisconsin–Madison Archives*

landowners. He learned that stewardship-minded private landowners have a different relationship with the land than do federal agents or the people who use public land. In his later years, Leopold became adamant that the highest form of conservation is achieved by the private landowner motivated by strong land stewardship ethics.

The Leopold Farm

In 1935, while a professor at the University of Wisconsin, Leopold purchased what he referred to as a "worn out farm" in the "sand counties" region of South Central Wisconsin. It was here that Leopold developed a close relationship with a specific piece of land. Long before, he had developed attachments to other tracts of land but not with the same level of detail and passion. There is something about owning a specific piece of land that is different from visiting lands owned by someone else. The pride of ownership and the responsibilities

Part of the Leopold family in 1938 at the Shack, a converted chicken coop used as the farm headquarters. Aldo and Estella (*at left*), youngest child Estella Jr. at age eleven (*in doorway*), and oldest child Starker at age twenty-five. *Courtesy of the Aldo Leopold Foundation and University of Wisconsin–Madison Archives*

of stewardship were things he discovered only after becoming a landowner. It was not a farm in the sense of raising crops or livestock. It had once been actively farmed but was in badly damaged condition, abandoned and needed time, loving attention, and good management to recover.

Leopold, his wife, and their children made countless trips to the farm on weekends, and it seemed to be the glue that kept the family bonded. Together they planted trees, restored grasslands, hunted, hiked, banded birds, and celebrated special occasions. Here is where they learned what it means to love the land and to become responsible land stewards. It was here that Leopold contemplated and wrote many of his finest essays. The farm was his primary outdoor laboratory and where he brought his students to learn the practical side of land management, including both successes and failures.

Leopold in Texas

Leopold made only two noteworthy trips to Texas during his career, one at the very beginning and one near the end. In spring of 1909 at the end of his final college semester, Leopold and his Yale classmates were sent to the deep woods of East Texas at Doucette for a ten-week training camp in preparation for employment with the Forest Service.[1] Despite the mosquitoes, rattlesnakes, and wild pigs, Leopold greatly enjoyed this time in a new environment, and it reinforced his desire to work as a forester.

Thirty-nine years later, in February 1948, Leopold made his final visit to Texas at the invitation of Robert Kleberg, president of King Ranch. Leopold's reputation had grown, and he was then regarded as one of the premier wildlife management experts in the country. He had developed a close professional relationship with most of the nation's top ecologists and wildlife professionals, including Val Lehmann, wildlife biologist of the King Ranch, who no doubt initiated the invitation. The purpose of the visit was for a mutual exchange of practical wildlife management information.

Leopold was very impressed by what he saw on the King Ranch, although he gently challenged their aggressive predator control measures. On the King Ranch, Leopold observed the integration of large-scale commercial livestock production alongside natural wildlife diversity and active wildlife management.

Afterward, Leopold wrote to Kleberg, "The trip over King Ranch was the highlight of my year and I want to thank you for your generous hospitality. I shall never forget it."[2] That was high praise from a man not given to flattery or exaggeration.

In later correspondence, Leopold wrote, "The King Ranch is one of the best jobs of wildlife restoration on the continent and has almost unparalleled opportunities for both management and research. I hope and pray that the King Ranch will never be broken up; it is a gem, the value of which few can appreciate."[3]

Leopold was so impressed by what he saw that he planned to incorporate it into his wildlife management classes at the University of Wisconsin. He wrote to Lehmann, "I am hoping to put together a lecture on the ecology and management of the King Ranch. I am interested in the livestock and range management as well as game."[4] Leopold had long believed that livestock management and wildlife management needed to be considered together, not separately. This is still an important lesson for all—that agriculture and wildlife management goes hand in hand.

Even though none of Leopold's actual life work was conducted in Texas, the principles he discovered and most of the basic tenets of ecology, conservation, and stewardship are universal.

His Writing

Leopold was an exceptionally gifted writer, and he sharpened his gift by constant use. His informal writing began as a young boy keeping a diary of bird observations and behavior. He kept the lifelong habit of recording nearly every hunting, fishing, birding, and camping trip, often in great detail, and these journals have fortunately been preserved by the University of Wisconsin.

Leopold kept careful records of every hunting trip, including the kinds and numbers of birds taken, the number of shots fired, and the sex, age, and weight of every bird. *Courtesy of the Aldo Leopold Foundation and University of Wisconsin–Madison Archives*

Leopold with dog Flick in about 1947. Leopold enjoyed bird study just as much as hunting and in his later years could often be seen with binoculars rather than shotgun. The basic principles of wildlife management that he discovered apply equally well to songbirds as to big game. *Courtesy of the Aldo Leopold Foundation and University of Wisconsin–Madison Archives*

Leopold's published writing started in 1911 and ended abruptly in 1948 with his unexpected death. He wrote or coauthored about five hundred articles, demonstrating vast knowledge and expertise in a wide array of land management issues.

Leopold's first book, *Game Management*, was published in 1933. This was the first book dedicated to the new science of wildlife management, and the principles he described are still used today. It was here that Leopold proposed his now famous thesis of game management—that wildlife habitats can be restored by the creative use of the same tools that previously had destroyed them. That thesis has become the credo of countless land managers across the country and perhaps the world.

Leopold's other book, for which he is most remembered, *A Sand County Almanac*, is primarily a compilation of essays and articles previously published, but it also contained some new material. He was putting the finishing touches on it when he died in 1948 of a heart attack while fighting a fire on his neighbor's farm. His son Luna completed the final editing, and the book was published in 1949 and since has been republished in various formats. However, much of Leopold's best writing is not found in either of these books but rather in articles he wrote. Many of these are reprinted or excerpted in anthologies and books about Leopold, as evident in the notes at the end of this book.

While *Game Management* is a book of science, *A Sand County Almanac* is a book of philosophy, specifically humankind's relationship to the land. Leopold was equally qualified to address both ends of the land management spectrum—the scientific methodology and the ethical motives, using the full capacities of mind and heart.

Who Is Aldo Leopold?

Who is Aldo Leopold? The answer is complex, just as the man was complex. For the private landowner, Leopold stimulates a deep desire to better understand and better care for one's piece of land, regardless of size. For the hunter and fisher, he instills the development of higher and higher ethical standards of sportsmanship and greater and greater outdoor skills. For the birder, he encourages not merely the identification of birds but also the study of their ecology, habitat, and behavior. For those engaged in wildlife management, he is still universally regarded as the father of that profession.

Leopold was a prolific and inspiring writer. Many of his finest essays were penned in this setting at his farm with his dog Gus nearby. *Courtesy of the Aldo Leopold Foundation and University of Wisconsin–Madison Archives*

For the student, whether beginning or advanced, Leopold kindles the curiosity to observe, question, contemplate, and understand the natural world. For the teacher or professor, he stresses the interrelationships of nature and that all of the natural sciences must be linked together, not just studied individually. For the parent raising children, he demonstrates the great and lasting value of spending much time together outdoors in nature and then allowing children to increasingly explore and learn on their own.

For the environmentalist, Leopold reminds us that humans are an integral part of nature and that nature produces a surplus of benefits that can be forever produced and used by people without exhaustion if managed wisely.

He challenges nature lovers not just to see the obvious or the lovely but also to probe, investigate, appreciate, and comprehend the interwoven and sometimes brutal nature of the land. Leopold urges all of these diverse groups to consider more deeply our relationship to the land and how it is managed, cared for, and sustained.

Fred Greeley (*left*), one of Leopold's graduate students, with Leopold after a woodcock hunt with bird dog Gus in 1946. Leopold invested himself in his students in the classroom and the field. *Courtesy of the Aldo Leopold Foundation and University of Wisconsin–Madison Archives*

2

LESSONS IN
ECOLOGY

"Ecology" simply means the study of the interrelationships and connections among and between all parts of the environment, including humans. Ecology is the foundational science that should govern our management of the land. If we ignore or misapply ecological principles, we will suffer consequences both presently and in the future. As we apply and integrate sound ecology-based principles into our management, we reap many tangible and intangible benefits, sometimes called ecosystem services.

The science of ecology integrates many closely associated disciplines, including soil science, botany, zoology, chemistry, hydrology, agricultural science, and wildlife management. The nine lessons below provide a groundwork of practical ecological thinking needed to successfully manage and sustain the land.

Balance of Nature

The "balance of nature" is a mental image for land and life which grew up before and during the transition to ecological thought. . . . To the ecological mind, balance of nature has merits and also defects. Its merits are that it conceives of a collective total, that it imputes some utility to all species, and that it implies oscillations when balance is disturbed. Its defects are that there is only one point at which balance occurs and that balance is normally static.[1]

Aldo Leopold, 1939

Most children learn about the balance of nature concept in elementary school, and it can be a good start toward understanding the natural world. My mom taught me about the balance of nature before I ever learned it in school. She suggested that it would make a good science fair project, so with her help I demonstrated a balanced ecological system. It consisted of a small aquarium with a few fish and a few plants. Fish need oxygen, and plants need carbon dioxide. Sunlight provided the energy for plants to generate oxygen through photosynthesis, while the fish consumed the oxygen and gave back carbon dioxide to sustain the plants. Microscopic algae provided food for the small fish, and the algae in turn were nourished by the waste from the fish. The system, although simple and artificial, was balanced and helped teach me something of how nature works.

The simplest kind of balance can be visualized as two children on a seesaw. Even if two are of equal weight, the balance is one of constant movement, not a stationary balance. As Leopold points out, the only time a true balance exists is the brief moment in each cycle where the beam is horizontal. So, in reality, dynamic imbalance is much more the norm than exact balance, and this is also true in nature.

A more accurate analogy would be to think of nature as an intricate machine with many spinning, oscillating, and meshing parts. In this case, balance occurs as long as the interconnected parts continue to spin, oscillate, and mesh as designed. If a large flywheel is added to the machine, it provides the momentum and gyroscopic effect to keep the machine running and balanced even during minor disruptions. While the analogy is still crude, it is perhaps more representative of a natural system.

Humans are obviously a part of the natural ecological machine and can either help maintain a balance or hinder and upset the balance. Leopold has said that proper human management lubricates the mechanisms of nature. Taking humans out of the equation can be one of the ways that the balance is upset. It can be argued that nature needs humans just as much as humans need nature.

If we tinker with the ecological machine too extensively or damage the parts, it gets out of balance or worse. It may continue to run but inefficiently and with a lack of power. The V-6 engine in my pickup will run on three cylinders—but very poorly.

A balanced natural system is one in which all necessary parts are retained and their movements stay within normal limits. Normal disturbances are easily accommodated in a balanced natural system, and certain disturbances such as fire or grazing are sometimes necessary or beneficial to sustain the natural dynamic balance.

When more severe disturbances occur, the system becomes temporarily impaired. Think of the 2011 Texas drought, which was extremely damaging to plant life. Damage to vegetation resulted in damage to soils and water resources and harm to animals and people. When the balance is upset, there is always a domino effect. However, even after extreme or abnormal disturbances, nature sooner or later begins to repair and rebalance damaged systems.

Weeds, brush, pests, parasites, and diseases are things that agriculture has tried to eradicate or control, in many cases for legitimate reason, but sometimes we have gone too far in trying to get rid of things that are part of the balance. In other cases, we tolerate an unnatural overabundance, which can also upset the balance. This includes too many livestock, excessive wildlife populations, or invasive plant species that dominate at the expense of natural diversity.

The balance of nature is a useful concept even if it does not fully explain the intricacies of ecological systems. For landowners, the key is to continually strive to manage in sync with the natural balance rather than trying to overpower nature. When upsets do occur, we can marvel at and cooperate with the ability of nature to rebalance.

Interdependence

The greatest single gain since 1930 lies . . . in the idea that resources are interdependent. We knew then that you can't have healthy fish in sick waters.
We knew something of the interdependence of animals and forests. But the idea
of sick soils undermining the health of the whole organic structure had not been
born. We had a vague notion that game and predators might be interdependent,
but we lacked details as to why and how.[2]

Aldo Leopold, 1939

The concepts expressed by Leopold in 1939 seem elementary to us now, but
these were new ideas at the time. Clearcutting vast forestlands, overgrazing
ranges, over-plowing prairies, and eradicating predators were the norm
during the 1930s. No wonder the land and water was sick.

The 1930s were difficult times in the United States. It was the Dust Bowl
decade dominated by drought, economic depression, and rumors of war.
Between severe droughts were the disastrous floods of 1935. Natural resources
were being ravaged simultaneously by nature and by poor management.

Our forebears who were farming and ranching at that time did not know
much about ecology or conservation or the interrelationships of soil, water,
crops, pastures, forests, livestock, wildlife, fish, and people. They did the best
they knew how but lacked a basic understanding of how the land works.

We have come a long way since the 1930s. We now acknowledge that all
parts of the land are connected and interwoven, even if we do not fully understand it and even if we do not always practice in accord with what we know.

Leopold once wrote, "To keep every cog and wheel is the first precaution
of intelligent tinkering."[3] We have all been in the situation where we hastily
take something apart, attempting to fix it, but on reassembly we find that we
have lost some of the parts, or, if we kept all the parts, we forget how they fit
together. Our land management can be likened to tinkering with a machine—
adjusting, lubricating, repairing, and sometimes overhauling. It should only
be done thoughtfully, carefully, skillfully, and with the recognition that all
the parts are necessary and interdependent.

The aftermath of a catastrophic wildfire on a ranch in the Texas Panhandle. Wildfire is one of the most devastating and demoralizing things that can happen. It affects people, livestock, wildlife, soils, plants, and waters in ways that are difficult to describe in words. Yet such disturbances are part of the natural balance, and the land will recover with good care and rainfall.

One of the ways that we attempt to understand the mutually dependent ecological relationship of people and natural resources is called holistic management. It is the realization that all parts and processes of the land organism are needed and must be integrated for functional and efficient operation. But holistic management does not primarily look at the individual parts; the focus is on seeking to understand the whole, which is always greater than the sum of the parts.

When our management of farms, ranches, and forests is guided by the underlying ecological principle of interdependence, good things tend to happen. Land is more productive and therefore has the capacity to be more profitable. Soils are healthier and hold more water and nutrients. Watersheds function better to process and protect the waters. Rivers and creeks have more sustained flows, and water quality is better. Wildlife populations are more stable, diverse, and strong. Biological diversity and carbon storage is enhanced. Land that has been damaged begins to heal.

When land is functioning properly with the parts in sync, both society and landowners benefit. There will still be droughts, floods, disease, and wildfire, and these setbacks can be extreme, but healthy land bounces back better than sick land.

Only later in life did Leopold discover the relationship of predators and big game. That knowledge was not intuitive and did not come naturally. He had to learn from his mistakes. In his early years, Leopold adamantly promoted the eradication of predators in order to produce more deer. It worked, but it worked too well. In the absence of predators, the mule deer population exploded to the point that the deer ruined the habitat and starved. He gradually changed his position on predator eradication once he realized the interdependence of game, predators, and habitat. We learn some lessons the hard way.

Interdependence is the opposite of independence. Many landowners embrace the spirit of rugged independence, but when it comes to managing the complexities of the land, the recognition of interdependence is vital.

One ranching family that embraces these concepts states their management theory this way: "We realize that the decisions we make on this ranch have a rippling effect." Let us always remember that everything we do affects something else and somebody else.

A Boring Bush

I want to plant some thoughts about a bush. It is called bog-birch. I select it because it is such a mousy, unobtrusive, inconspicuous, uninteresting little bush. . . . It bears no flower that you would recognize as such, no fruit which bird or beast could eat. It doesn't grow into a tree which you could use. It does no harm, no good; it doesn't even turn color in the fall. Altogether it is the complete biological bore.

But is it? Once I was following the snow tracks of some starving deer. The tracks led from one bog-birch to another; the browsed tips showed that the deer were living on it, to the exclusion of scores of other kinds of bushes. Once in a blizzard I saw a flock of sharptail grouse, unable to find their usual grain or weed seeds, eating bog birch buds. They were fat. . . .

It appears, then, that our little nonentity, the bog-birch, is important after all.[4]

Aldo Leopold, 1939

Bog birch does not grow within a thousand miles of Texas, but Leopold's account provides an important ecological lesson that applies to every farm, ranch, forest, and wetland. The lesson is not just about obscure shrubs; it is also about appreciating species we do not yet understand. Texas has many such bushes, which are often lumped into the category of "brush." These obscure secondary bushes are seldom abundant enough to be considered a problem, and they do no harm. Yet to the casual observer they are worthless plants that could be replaced by something of value.

A well-respected range management specialist once likened algerita to a wart on the nose—nothing more than an ugly blemish to be removed. As one who was trained in ecology, he should have been able to see the value of the lowly, spiny algerita, but too often we only see what we are programmed to see. Since it takes up space, and because it is relatively poor feed for livestock, the specialist viewed it as a good-for-nothing bush. A closer evaluation of algerita with an open mind would have revealed considerable seasonal value for deer and songbirds and as a prime nursery shrub to protect other palatable and desirable species.

Baccharis is another example. Also called poverty weed or Roosevelt weed, it is an unattractive scruffy bush that frequently grows in gravelly creek and river bottoms. The most common question about baccharis is "How do I kill it?" Few have bothered to discover what value it brings.

Baccharis is a superb gravel bar pioneer, able to grow in a harsh environment. In riparian areas, baccharis slows the velocity of floodwaters, reducing erosive forces. The multibranched shrub catches fine leaf and twig debris which in turn catches soil and seed. Other important species such as switchgrass and American sycamore often get their start in the protected debris-catching zone created by baccharis. Baccharis is like a scab on a flesh wound—an important step in the healing process of scoured gravel bars. It is also happens to be a rich source of nectar for fall butterflies.

This lesson is not just about bushes. I once asked my friend Frank Price who ranches west of Sterling City, Texas, what he thought of the grazing value of slim tridens, a common grass on many Texas ranches. After a long pause, he gave his usual thoughtful answer, "I don't really know, but I'm glad I have it."

Frank could not say for sure what the grazing value was, but he recognized that it was a native grass that belonged and had a niche on his ranch. He had learned not to judge a grass on its forage value alone, something that many of us need to understand.

As it turns out, slim tridens seeds provide good food for quail, turkey, and other ground-feeding birds. It holds topsoil in place and adds organic matter to the soil. While it is not a cow's favorite grass, they do graze it at certain times, and it has other values that may be just as important.

Javelina bush, Mexican buckeye, mariola, amargoso, spiny aster, junco (allthorn), shrubby blue sage, paloverde, catclaw mimosa, lechuguilla, wafer ash, and rabbitbrush are just a few of the many plants that we too often place into this category of useless, boring bushes. Yet, like bog birch, each has its own story and its own value.

This lesson recalls Ralph Waldo Emerson's definition of weeds as plants whose virtues have yet to be discovered.[5] The ecologically minded landowner or manager will be wise to adopt and cultivate a greater appreciation of the many obscure species we do not yet understand.

Soil and Water

Soil and water are not two organic systems, but one. Both are organs of a single landscape; a derangement in either affects the health of both.[6]

Aldo Leopold, 1941

An article or a speech about soil and water may sound uninteresting to the average citizen, but to people of the land, soil and water are everything—the basis of all life. Leopold states an important ecological lesson; what happens to the soil affects the water, and taking care of the soil is the best way to take care of the water.

When farmers, ranchers, and forest owners take proper care of the soil, good things happen. Springs, creeks, rivers, wetlands, and lakes will be cleaner, better sustained over time, and abounding with life. Healthy soil also results in more drought resilience and less severe and less frequent flooding. Unhealthy soil results in a more arid landscape, increased runoff, increased erosion, poor water quality, and reduced productivity.

In 1914 Ray Dickson became the superintendent of the Texas Agricultural Experiment Station in Spur, Texas. Those were times of economic hardship and drought for landowners in West Texas. Farmers and ranchers were desperate, pleading with God to send rain. During these hard times Dickson reportedly offered some practical hydrologic advice: "Don't pray for rain if you can't take care of what you get." He understood that the condition of the soil is what determines how much rain soaks in and how much is captured and stored. Dickson's advice is still valid today.

Leopold understood that soil is not just a mixture of sand, silt, and clay. These mineral particles comprise only about half of the volume of healthy soil. The other half is pore space, teeming with billions of microorganisms carrying out the complex biological and chemical processes that make soil productive and wholesome. The pore space is also the soil's vascular and respiratory system, allowing water, nutrients, and air to move through it.

Organic matter, the key component of healthy soil, is also housed in this pore space. These organic particles of decomposed plant and animal matter

Several major rivers originate in the karst limestone water catchments of the Edwards Plateau. Private landowners are increasingly aware of the vital connection between land and water and the importance of responsible stewardship.

are what sustain the intricate food webs of microorganisms, which in turn keep the soil strong and rich. A modest organic matter content of 2 percent translates into twenty tons of soil organic matter per acre. For every small increase in organic matter, there is an exponentially large increase in water-holding capacity, fertility, and productivity.

Over twenty-four hundred years ago, Plato described what must have been a perfectly functioning water cycle in ancient Greece, noting the importance of healthy soil. In *The Republic* he noted that the mountains and hills were well-covered with soil that provided an abundance of timber, and rich grasslands provided abundant forage for livestock. He observed how the land soaked up and stored the rainfall so that springs were plentiful and streams were flowing. He concluded his description by remarking that the land was cultivated by "true husbandmen" and reemphasized the richness of the soil and the abundance of water.

The basic principle of soil and water conservation is straightforward: keep the soil well-covered in all seasons with a wide variety of living plants and a layer of decomposing plant litter. Violating this principle results in degradation of the soil and water.

The water supplies and water quality of Texas largely depend on private agricultural and wildlife management lands. When those lands remain intact and well taken care of, it will help ensure healthy and abundant waters as well as food, wildlife, and recreation for future Texans.

A Fountain of Energy

Land . . . is not merely soil; it is a fountain of energy flowing through a circuit of soils, plants and animals.[7]

Aldo Leopold, 1939

When we think of energy, we traditionally think of hydrocarbons such as oil, natural gas, and coal, or the electricity that is generated from these fossil fuels. Nowadays we also think of wind farms and solar energy. We think of the pipelines and powerlines that crisscross the land transporting energy from where it is produced to where it is used.

We usually do not think about the land itself as a component in the flow of energy, although this is the planet's primary means of energy production and transfer. Every young student learns that sunlight energy striking green leaves is transformed into a different kind of energy by the process of photosynthesis. Plant energy is what sustains all life on earth, either directly or indirectly.

Scientists calculate that on average about 126 watts of sunlight energy per square foot strikes the earth during daylight hours. On one acre with an average of ten hours of sunlight per day this would be enough raw energy to power 1,890 homes or the energy equivalent of 1,629 gallons of gasoline each day. It is an almost unbelievable amount of raw energy, the same energy that is available to grow trees, grass, wildflowers, and crops.

It is interesting to consider that the fossil fuels we use each day are also the result of this massive prehistoric transfer of sunlight energy to plant energy. Eons later, that sunlight and plant energy, now in a different form, is stored in the ground for human discovery and use.

Our management of the land determines in large part to what degree sunlight energy is put to beneficial use. Energy conservation on the farm or ranch is a lot more than being careful with fuel and electricity consumption. To the land manager, energy efficiency can be viewed as managing every square foot to maximize the conversion of the sun's energy through beneficial plants. Although the flow of energy is one of the most complex aspects of ecology, the practical application is much simpler.

The biggest wastage of energy on the farm or ranch is sunlight striking bare ground. In the summer, bare soil absorbs solar energy and heats up to 130 or 140 degrees —hot enough to kill or suppress the growth of essential soil microbes. This in turn hinders soil health, infiltration rates, and water-holding capacity and dramatically reduces productivity. While we may cook our steaks to 130 degrees, we should not be cooking the soil. An overheated soil, just like an overheated motor, has impaired productivity and is a sign that things are terribly out of order.

In the grand circuit of the land, the energy in plants is transferred to the animals that consume plants, whether livestock, deer, rodents, or insects. In some cases, the food chain is simple, as in the case of cattle eating grass, and in other cases it is more complex, with several links.

At each step in the food chain energy is lost, and, at the end of the chain, decomposition breaks down the dead plant or animal. The resulting energy (carbon) is cycled back into the soil and the atmosphere to fuel next-generation plant growth. It is difficult to comprehend how perfectly the system operates when it is working properly.

On ranches, there are several ways to ensure the efficient conversion of sunlight energy to plant and animal energy. Green plants should be growing all throughout the year, even in the winter. This means that cool-season plants and evergreens should be mixed with warm-season species to ensure year-round photosynthesis.

A healthy energy flow also requires a diversity of plants, which enables the most effective capture and conversion of sunlight. Deep-rooted plants, including many woody species, are able to remain photosynthetically active even in drought. Short-lived annuals provide a quick but temporary burst of green cover while the slower-growing perennials provide more stable, long-term cover. Every different type of plant has its own niche in the energy circuit.

Animal diversity is also needed for the efficient transfer of energy. Soil organisms, insects, rodents, small mammals, larger herbivores, birds, predators, and decomposers are all needed to keep the circuit humming. A gap or constriction at any stage interrupts or weakens the flow of energy through the system.

Landowners tweak the controls of the terrestrial energy cycle with grazing, fire, brush control, timber management, farming practice, and hunting. By increasing or decreasing the pressure and timing of these adjustments, humans become a key part of the energy circuit. We admire the natural ecological balance that has been created, and we humbly acknowledge our responsibility of caring for it and managing it for beneficial purposes.

Dark and Bloody Ground

We have not yet learned to think in terms of small cogs and wheels. Look at our own back yard: at the prairies of Iowa and southern Wisconsin. What is the most valuable part of the prairie? The fat black soil, the chernozem. Who built the chernozem? The black prairie was built by the prairie plants, a hundred distinctive species of grasses, herbs, and shrubs; by the prairie fungi, insects, and bacteria; by the prairie mammals and birds, all interlocked in one humming community of co-operations and competitions, one biota. This biota, through ten thousand years of living and dying, burning and growing, preying and fleeing, freezing and thawing, built that dark and bloody ground we call prairie.[8]

Aldo Leopold, 1938

Leopold asks: "What is the most valuable part of the prairie?" And his answer is not too surprising. The most important part of the prairie is not the prairie chicken, bison, or pronghorn antelope. It is not the meadowlark or the prairie dog. It is not the pristine prairie grasses or wildflowers. Leopold proclaims that the most valuable part of the prairie is the soil, and then he elaborates on the incredible ecological process of how prairie soils are formed and sustained.

He refers specifically to a type of soil known as chernozem. Chernozem soils are among the richest and most productive soils in the world and are found in parts of Russia, Canada, and the midwestern United States. Texas has no true chernozem, but the deep dark soils of the Blackland Prairies share many of the same characteristics.

These rich prairie soils are characterized by very high organic matter content. Every gardener knows the magic of organic matter. Organic matter, also called organic carbon, is what sustains life in the soil. In grassland, organic material comes primarily from the normal annual turnover and decomposition of grass and forb roots and from the litter layer that forms surface mulch that is gradually broken down into compost by soil organisms.

At the microscopic level, healthy soil contains countless fungi and bacteria, which process raw organic material. Termites and earthworms also help break down organic material into small particles. Nematodes and protozoa consume the fungi and bacteria. Larger organisms such as mites, springtails, sow bugs, ants, and millipedes consume the smaller organisms in turn, creating an intricate underground web of life. The constant dying of these organisms is why Leopold calls it "dark and bloody ground."

When the life in the soil is abundant and balanced, the soil is rich, fertile, mellow, and highly porous. This porosity allows the rapid movement of water and air into and out of the soil. Such a soil has exceptional capacity to hold water and nutrients and make them available for plants. Leopold recognized the soil as a renewable resource. If managed properly, we reap the benefits in perpetuity. If land is mismanaged or abused, the productivity of the soil and its ability to sustain life is greatly reduced.

Leopold constantly encouraged people to think about the small cogs and wheels that make up the inner workings of the land. He compared the land to a complex machine with many parts, each working in synchronization with other parts. When even a small part is missing or broken, the whole machine falters.

Today in conservation circles there is renewed excitement about soil health. Modern-day conservationists and soil ecologists are rediscovering what Leopold described many years ago. The chemistry, biology, and hydrology that take place in the soil comprise one of the most complex, remarkable, and least understood parts of nature. Our supplies of food, water, wood, and wildlife depend on healthy, functional soils. Agricultural production can be done in ways that maintain or restore the health and vitality of the soil. As depleted soils are restored through good conservation practice, agricultural productivity increase, and healthier waters, more wildlife, and other natural benefits will result.

Our role as soil stewards is fairly straightforward in theory. Keeping the soil covered by a wide variety of living plants and plentiful decomposing plant litter in all seasons is the key to maintaining a healthy, living, productive soil. Doing this in real life is not so simple and requires skill and dedication.

Jon Taggart, who raises grass-fed beef in the Blackland Prairies of Texas, weathered the historic and devastating drought of 2011 without having to liquidate his herds or reduce his stocking rate. When asked how this was possible, he replied, "We have been preparing for this drought for the last twenty-five years." By grazing conservatively, always keeping a cover of grass, rotating the livestock, and giving each pasture ample time for rest and recovery, he had been increasing the root systems of his grasses and simultaneously restoring and improving the health and vitality of the soils. The soils, covered with a layer of protective plant litter, were being gradually enriched with increasing organic matter and soil-building microbes. The result was a soil that held every drop of water that fell and grasses with a deep, healthy root system to make the most efficient use of whatever moisture was available.

Ecologically minded landowners are keenly aware of the intricately interwoven nature of the land and the complex mechanisms that keep the land in good working order. As land stewards and conservationists, we can admire and appreciate the natural world even though we do not fully understand its complexities. May we all strive to better understand and better care for the resources entrusted to us—the soil, the water, the plants, and the animals.

In the Chinati Mountains of West Texas a good covering of native grass and sotol keeps the thin soils covered and well protected, allowing rainfall to soak in, then coming out elsewhere as springs.

Ecological Face Powder

Today the honey-colored hills that flank the northwestern mountains derive their hue not from the rich and useful bunchgrass and wheatgrass which once covered them, but from the inferior cheat which has replaced these native grasses. The motorist who exclaims about the flowing contours that lead his eye upward to far summits is unaware of this substitution. It does not occur to him that hills, too, cover ruined complexions with ecological face powder.

The cause of the substitution is overgrazing. When the too-great herds and flocks chewed and trampled the hide off the foothills, something had to cover the raw eroding earth. Cheat did.[9]

Aldo Leopold, 1941

Cheatgrass (*Bromus tectorum*) is one of the many species of exotic grasses that now monopolize untold millions of acres of the American landscape to the detriment of native grasslands. Fortunately, Texas does not have severe infestations of this species, but we have other nonnative grasses that often dominate our landscapes. To the untrained eye, the hillsides, prairies, savannas, and shrublands occupied by these grasses may look good and healthy. To livestock producers, these exotic grasses provide forage, protect the soil from erosion, and provide some legitimate ecological function. However, to the trained eye, these grasslands dominated by nonnative species represent sick landscapes and diminished wildlife habitat.

Face powder does not enhance beauty, it merely hides defects. Likewise, exotic grasses superficially cover up defective landscapes on which the native plants and healthy soil have been damaged. Native plant enthusiasts often ascribe negative and even hateful emotions to exotic plants, but this is faulty thinking. In most cases, these plants are merely filling the niches we have created. Cheatgrass and other exotic grasses do not destroy the beauty of healthy native grasslands; rather, they colonize, establish, and increase in places that have already been damaged and degraded by other causes.

There are two ways that these exotic grasses find their way on to our farms, ranches, and roadsides: either they sneak in uninvited, as was the case with

These honey-colored hills of the Edwards Plateau may look beautiful to the untrained eye. But overzealous clearing and the establishment of nonnative KR bluestem indicates a sick landscape.

cheatgrass, or they are intentionally planted. In Texas most of our troublesome exotic grasses have been intentionally imported and planted for their livestock forage value. Once established, some of these grasses have spread exponentially. We are now reaping the consequences of what we have sowed, and this is an important ecological lesson to remember.

We will never rid Texas of exotic invasive grasses, but there are at least three ways to diminish their influence and spread. The first and most intuitive approach is to stop planting them. Some landowners continue to plant these so-called "improved grasses" because they are relatively cheap, productive, easy to establish, and easy to manage. This practice will probably continue as long as federal conservation agencies encourage and provide incentive for this practice.

The second approach is to manage rangeland in a way that sustains and improves native plant diversity. In this way, the ground will remain covered by native vegetation, thus reducing the opportunity for encroachment and invasion. This is the most practical method to manage and minimize these grasses. The third approach is to kill or suppress the nonnative grasses and to replace them with an appropriate mix of native species. This is much more difficult than it sounds and is presently only feasible on a small scale and with mixed results.

Leopold stated that overgrazing was the principal cause of the widespread invasion of cheatgrass. In Leopold's day, overgrazing was so widespread that it was considered the norm. Heavy grazing gradually removed the natural cover of protective native grass until the ground was barren and naked, leaving it vulnerable to nonnative weeds and grasses.

One of the chief ecological truths is that nature hates nakedness. Land is meant to be covered by plants, and a cover of exotic plants is better than no cover at all. The era of rampant and severe overgrazing was a dark and damaging time in natural resource management, and it caused great harm. We continue to see the effects today.

Fast-forward to the present, and the problem of overgrazing is much less prevalent than in Leopold's time. Across the entire country, grazing management has improved markedly in the past several decades. Some of this is due to ranching economics and recurring drought that has resulted in greatly reduced livestock numbers. Some can be traced to an increasing interest in wildlife management, which is closely tied to good grazing practices. More importantly, much of this improvement can also be attributed to an increasing and deepened commitment of responsible land stewardship among landowners.

One of the important principles of ecology and land management is the law of unintended consequences. These exotic grasses that sometimes dominate the landscape were imported with good intentions, for beneficial purposes. No one suspected that these grasses would one day occupy millions of acres, damaging native plant diversity and wildlife habitat. The side effects were not anticipated.

Few things in land management and conservation are black-and-white. Most activities have benefits as well as potential risks. Exotic grasses have some valid ecological and agricultural use, as noted by Leopold, but these benefits must be weighed against the potential long-term negative impacts. No one can accurately predict the future effects of our actions, but land stewards are those who carefully and prudently think through land management options and try to anticipate the rippling effect to the land, their neighbors, and future generations.

Leopold would be gratified today to see such a large and growing movement of people dedicated to restoring native grasslands. He lamented that no such movement existed in his day. Today Leopold would be smiling to find large, enthusiastic groups of landowners who take pride in being responsible stewards of native plants and animals.

A Healthy Land Organism

Conservation is a state of health in the land. The land consists of soil, water, plants, and animals, but health is more than a sufficiency of these components. It is a state of vigorous self-renewal in each of them, and in all collectively. Such collective functioning of interdependent parts for the maintenance of the whole is characteristic of an organism. In this sense land is an organism, and conservation deals with its functional integrity, or health.[10]

Aldo Leopold, 1944

Leopold's philosophy of conservation described here is far more complex than our traditional concept of conservation. For some landowners and government agencies, conservation is still defined largely by implementing certain practices. The list includes brush and weed control, reseeding, rotational grazing, water development, cross fencing, food-plot planting, prescribed burning, timber thinning, contour farming, planting of cover crops, and dozens of other practices. Applied correctly and in the right situation, these can all be beneficial practices, but according to Leopold this is not the essence of conservation. He says that conservation is really about healthy land, where all the components are vigorous, self-renewing, and functioning properly.

Leopold likens the health of the land to the health of a living organism. We can use the human body as an example. The human body consists of twelve organ systems, seventy-eight distinct organs, and thousands of integrated parts. There are 270 bones, 900 ligaments, 650 muscles, and 4,000 tendons. We each have about forty-six miles of nerves and sixty thousand miles of blood vessels. In addition, our bodies contain about as many nonhuman microbial cells as there are human cells, and these microbes help our bodies function properly. When these collective parts and systems are in working order, we are said to be healthy. A defect in any of the parts can cause the body to be out of order.

The things that can upset the health of the body are numerous—stress, strain, injury, infection, nutritional deficiency, obesity, substance abuse, genetic disorders, and many others. These include things we bring on ourselves and things that happen due to no fault of our own.

Likewise, there are things that upset the health of the land, and these include things we do as part of our management as well as things that just happen. Injury to the land organism can result from overgrazing, excessive deer or exotic (nonnative) ungulate populations, over-clearing of brush, excessive weed control, creating a monoculture of grass, burning under improper conditions, removal of predators, farming practices that cause erosion, or other forms of mismanagement. Damage can also occur from things over which we have little or no control, including drought, floods, hurricanes, disease, insect plagues, and other acts of God.

Despite the prevalence of sickness, both the human body and the land organism are resilient and will often bounce back naturally from sickness or injury. Many times the disorder will start to correct itself if we just stop doing what caused it. If the damage is severe enough, it may warrant an aggressive treatment. However, in many cases it is simply a matter of balance and moderation, such as consuming less sugar and getting more exercise. Tweaking our habits is often sufficient to improve the health of body—or land.

In his day, Leopold realized the tendency for land managers to try to fix the symptoms of sick land without discovering or treating the underlying cause. He said, "The practices we now call conservation are, to a large extent, local alleviations of biotic pain. They are necessary, but they must not be confused with cures. The art of land doctoring is being practiced with vigor, but the science of land health is yet to be born."[11]

This was true in Leopold's day, and to some extent it is still true today. However, in the last several decades the practical application of ecological principles of land health has made excellent progress. We now have a greater knowledge and experience to treat not only the symptoms of sick land but also to discover and treat some of the underlying causes.

Meredith Ellis, who ranches with her father on the G Bar C Ranch in North Texas, insists that one of the most important elements of successful land management is to first understand the principles that govern land health, not merely to apply practices. By focusing on the underlying principles rather than practices, the father-daughter ranching team is noticing an impressive rebalancing of the grasslands, causing the land to be more productive than ever before. With this emphasis, the ranch is producing sustainable beef, restoring biodiversity, providing wildlife habitat, filtering water, putting more carbon in the ground, and generating a profit.

There is much to be excited about in the world of conservation, as well as some continuing challenges. Leopold would probably be happy to see what is happening on private lands in Texas yet would encourage us all to keep working for healthy lands, waters, and wildlife.

Professors and Poets

There are men charged with the duty of examining the construction of the plants, animals, and soils which are the instruments of the great orchestra. These men are called professors. Each selects one instrument and spends his entire life taking it apart and describing its strings and sounding boards. This process of dismemberment is called research. The place for dismemberment is called a university.

A professor may pluck the strings of his own instrument, but never that of another, and if he listens for music, he must never admit it to his fellows or to his students. For all are restrained by an ironbound taboo which decrees that the construction of instruments is the domain of science, while the detection of harmony is the domain of poets.[12]

Aldo Leopold, 1940

The passage in the epigraph above, from *A Sand County Almanac*, was written following a Leopold trip to the Gavilan River region in the Sierra Madre Mountains of northern Mexico, about two hundred miles southwest of El Paso. Leopold made two hunting trips to this area, in 1936 and 1937, marveled at the natural balance that still existed there, and commented that the region was the picture of ecological health. The area was yet untouched by intensive agriculture or modern land management practices, and the "music" he experienced there and the harmonies he detected influenced much of his thinking in the years ahead. He used the metaphors of instruments, music, professors, and poets to speak about ecological relationships.

In this passage, Leopold highlighted what he saw as the overspecialization and isolation of scientists and scientific thinking. He likened the university and research system to those who skillfully study and play an instrument but are not able to hear the music created by fellow musicians in the orchestra.

Today's research efforts often share some of the same deficiencies and generate the same frustrations as Leopold noted. In our system of academic research we still find a great deal of isolation of disciplines. We have wildlife professors who cannot identify or discuss the significance of the common grasses. We have soil scientists who know little about plants and botanists who know little about soil. We have plant specialists who know nothing about the animals that depend on these plants. We have professors and researchers who dedicate themselves to studying deer, quail, turkey, waterfowl, songbirds, predators, fish, livestock, watersheds, hydrology, fire, brush control, grazing management, grasses, trees, and dozens of other specialties. In most cases these specialists are highly intelligent and excel in their respective niche. In too many cases these specialists have tunnel vision and are not able to see how their specialty meshes with the whole.

Leopold frequently spoke of the incredible complexity of the land organism. He realized that no one would ever be able to fully comprehend the intricacies of nature. However, successful professors must be able to explore, discover, and explain how their specialties intersect with other disciplines. Relatively few of the innumerable specialists in our universities and agencies are able to see the big picture.

Leopold himself was also a professor, the first professor of wildlife management in the United States. Leopold saw firsthand the compartmentalization of scientific research common in the university system. No doubt, he sought to teach his students and fellow professors to view and study the natural world with a wide-angle lens in order to be more integrated and less of a specialist.

The best natural resource research demands scientists who cooperate with those in other fields and understand the context of their specialty. Effective professors must not be soloists, no matter how talented they are, but instead must see themselves as members of an orchestra. They must seek to interact and integrate with other specialists and the real world.

Land managers and wildlife managers find themselves in a challenging position. They must base their actions and plans on the best science available, but they must also be able to perceive what science is not able to explain. Successful land managers must have the dispositions of both poets and professors. They must be able to understand the basic science behind their endeavors but also be able to detect, appreciate, and blend in with the complex harmonies of the land.

Reading scientific articles and research reports is beneficial for understanding some of the components of the land. While this is good, it is not enough. Getting out of the truck, sitting alone on a ridge, a river bottom, or a brush thicket and listening, watching, and using all of the senses to interpret the poetry and music of the land is also important. Using both the logic of the mind and intuitions of the heart is the task and the joy of the genuine land steward.

How do my actions and my plans harmonize and synchronize with what I know and what I perceive about the land and the health of the land? These are the great questions and the great pursuits of successful land management.

3

LESSONS IN
LAND ETHICS

Ethics are the inner values, motivations, and convictions that govern our actions and form the basis of our sense of right and wrong. Land ethics deals with why we do what we do regarding our treatment of the land. These lessons do not attempt to spell out what is ethically right or wrong for the individual since that must be established from within. The lessons below lay out the truth that our personal convictions and ethics are what drive our desire or lack of desire for land stewardship.

Our Relation to the Land

It is inconceivable to me that an ethical relation to land can exist without love, respect, and admiration for land, and a high regard for its value.[1]

Aldo Leopold, 1949

Aldo Leopold had much to say about man's relationship to the land. It is perhaps the subject he is most known for. Our relationship to the land dictates the way we treat the land, and, for many of today's landowners, managers, and nature enthusiasts, Leopold's assertion on ethical relations to the land states what they already believe.

Others besides Leopold have helped to define and refine what a proper relationship to the land looks like. The late John L. "Chip" Merrill is well known in Texas ranching and conservation circles and was the director and a primary instructor of the renowned Texas Christian University Ranch Management Program for many years. In this role he was instrumental in training his students to be conscientious land stewards as well as successful ranchers. His teachings, leadership, and example have now influenced several generations of landowners and conservationists. Merrill's words perfectly complement the ideas of Leopold, although the two men did not know each other. He said, "One cannot work closely with the land for very long without developing a deep respect and appreciation for its character and capabilities."[2]

The basic ingredients of stewardship ethics are described here by Leopold and Merrill. These two great conservationists, both of whom were private landowners, are in harmony in describing what it means to have an ethical relation to the land.

Merrill and Leopold each described man's relationship to the land in both emotional and pragmatic terms. A genuine land steward views the land from the heart and with the mind simultaneously—with both feeling and logic. Love of the land is primarily an emotional response, but this love always expresses itself in concrete ways and is demonstrated in how the land is treated. It is impossible to love the land yet abuse or neglect it.

Merrill described the sense of appreciation and respect that one develops when working closely with the land. "Working with the land" implies first-hand knowledge of the land, specifically how soil, water, plants, and animals operate together interdependently under the influence of human management. In this way, humans cooperate with the ecological laws of nature rather than overruling them.

Merrill spoke of the land having character and capability. This acknowledges that each piece of land is special and unique and must be managed and treated specifically in accordance with its capability. Responsible land ethics in concert with ecological understanding do not try to demand more from the land than it can naturally give. Stewardship works to improve, enhance, and restore the land in keeping with its potential but does not force it to exceed natural limitations.

Both Leopold and Merrill held a very high regard for the value of land. This was not just a high regard for the economic value of the land or its productivity but also a high regard for the intrinsic natural, ecological value of the land. Both men understand that when ecological functions are recognized and woven into the management of a farm or ranch, the land will be productive and the value sustained and enhanced in the long term.

The sum of these attitudes is a person having an ethical relationship to the land. The modern concept of land ethics was developed by Leopold and has been honed by Merrill and many others. Land ethics involves the inner conviction and inclination to be a responsible custodian of the land. Even when land is under the legal ownership of a person, a stewardship ethic insists that the owner is really just a caretaker of the land that is entrusted to his or her care.

Today in Texas, ranchlands, farmlands, forestlands, and wetlands are being increasingly managed and cared for according to the ethical ideals described by Leopold, Merrill, and other progressive conservationists. There is still a lot of work to be done. Not all landowners have an ethical relation to the land. In some cases, our land ethics are weak and unrefined, but Texas landowners are moving in the right direction. Texas is a better place because of the growing number of rural landowners who understand, respect, and appreciate the land and whose love for the land influences everyday land management decisions.

Limiting Our Freedom

An ethic, ecologically, is a limitation of freedom of action.[3]

Aldo Leopold, 1949

Most of us do not like to have limitations on our freedom, and we do not like being told what we should or should not do. Landowners especially tend to be an independent bunch and resist being told what to do.

For some readers, the title of this lesson may be a red flag. Anything that suggests a limit to our freedom is often met with suspicion if not outright rejection. Our country was built on the premise that we have many inherent and coveted freedoms, but even the most ardent freedom lover realizes that certain limitations are needed for the common good. Some freedoms are limited by law and some are restricted voluntarily. The limitations of freedom Leopold addressed are self-imposed limitations in the name of stewardship.

For the private landowner who is also a land steward, there is a well-developed sense of personal limitation that exists alongside the exercise of freedom and independence. The two ideas are not in opposition. Call them scruples, qualms, or convictions, these are the inner principles that guide our management, and they include a definite set of personal restrictions regarding what we may and may not do with our land.

There are many land management activities that are perfectly legal and that are commonly carried out by some landowners. However, other landowners may not feel the freedom to do some of these things for ecological or ethical reasons, even if they can see the possible benefits.

Consider these practices: It is still common for some landowners to plant aggressive exotic grasses that are known to spread and dominate entire landscapes. It is legal to cut down every tree, kill every stick of brush, and spray every broadleaf weed. It is legal to overgraze every blade of grass to the point of bare ground and erosion. It is legal to pump large volumes of irrigation water out of shallow alluvial wells, even when it causes the base flow of creeks and rivers to decline to a trickle or cease altogether.

It is legal to bait and then overharvest quail, turkey, or deer on your property. It is legal to set up a blind and feeder adjacent to your neighbor's pasture. It is legal to confine white-tail or exotics or both inside small enclosures far in excess of carrying capacity. It is legal to purchase, release, and hunt "shooter bucks" raised in captivity. It is legal to potshot quail or shoot ducks on the water.

These and many other things may not be inherently or universally wrong, but they may be very wrong for the individual and therefore self-prohibited. These self-imposed limitations are the basis of land ethics. For the land steward, not everything that is legal is ethical or permissible.

One of the earmarks of genuine land stewardship is the realization that whatever one does on one's own place has a rippling effect. What I do affects others and has side effects beyond my fence, so I am careful not to do things that might be detrimental to others or a poor example of sportsmanship or stewardship.

In his day, Leopold noted that there was no teaching about ecological ethics in universities or land management agencies, and this is still somewhat true today. The natural resource departments of most universities and our state and federal agencies still do not promote land ethics as the backbone of good land management. Herein may be one of the important roles of nongovernmental organizations such as Texas Wildlife Association that effectively promote sound stewardship ethics. Such organizations do not dictate individual ethics or promote the regulatory approach to resource management, but they do encourage thoughtful, voluntary consideration of ethical land management.

Leopold did not advocate for a heavy-handed government forcing limitations on freedom. He addressed strengthening the internal laws of conscience that exist in all of us and are more highly developed in the individual who loves the land and embraces the ethics we call stewardship.

We appreciate and celebrate the freedoms we have as Americans, including the right of owning land. But let us remember that there are certain freedoms that we voluntarily restrict in order to be good neighbors and good stewards of Texas lands, waters, and wildlife.

Landowner Obligation

The average citizen, especially the landowner, has an obligation to manage his land in the interest of the community as well as his own interest. . . . The nation needs, and has a right to expect the private landowner to use his land with foresight, skill and regard for the future.[4]

Aldo Leopold, 1946

Leopold makes a bold statement in asserting that landowners are obligated to manage their land for the benefit of the community, not just for their own personal benefit.

Many of today's farmers and ranchers will readily agree that it is desirable and commendable when landowners manage their land in a way that provides benefits to society, but not all agree that this is an obligation. In fact, by driving along country roads and looking across the fence it is easy to see that some landowners do not consider such management as their responsibility.

Leopold did not advocate the regulatory or forced approach to land management. Over and over again, throughout his life, he spoke well of the landowner who voluntarily took good care of the soil, water, plants, and animals. He spoke poorly of landowners who did not take care of their land. But he never spoke of forcing landowners to carry out good management. Leopold knew what we all know—that you cannot successfully mandate good ethics or good behavior.

How can something be both voluntary and obligatory at the same time? It seems contradictory. The only way this makes sense is through the lens of genuine stewardship, whereby the landowner voluntarily obligates themself to care for land.

Unfortunately, land stewardship is sometimes misunderstood simply as a checklist list of practices or accomplishments. But, properly understood,

The Canadian River meanders back and forth across its wide valley in the Panhandle of Texas. Stewardship-minded landowners seek to understand the natural dynamics and voluntarily obligate themselves to care for the land entrusted to them.

stewardship is the innate love and respect for the land that in turn compels and inspires owners to study, understand, and carefully manage the land that has been entrusted to them.

A significant and growing number of Texas landowners do manage their land in ways that benefit others. They do this not to accrue accolades or financial gain but because they consider it the right and responsible thing to do. For these landowners, it is a self-imposed obligation, not an option. These societal benefits that result from well-managed land are referred to as ecosystem services and include water quality, sustained flows, aquifer recharge, carbon absorption, wildlife and pollinator habitat, and aesthetic values.

Landowners may or may not consciously think of themselves as providing these services; these are the natural by-product of taking care of the land. For example, when deer habitat is properly managed, it is also good for songbirds and many other species. When grassland cover is maintained by proper grazing, watershed conditions are protected and carbon is stored in the soil.

What do you think? Is there a moral or ethical obligation for landowners to manage their land in a way that benefits others? Is this something that society has the right to expect? Should society provide some form of incentive or financial benefit for landowners who carry out this kind of management? Should there be any form of disgrace or disincentive for landowners who do not manage their land well?

These are not easy questions. They cause us to think more deeply and consider the bigger picture of landownership dynamics, including both the rights and responsibilities of landowners. As Texas grows more urban and more disconnected from the land, these questions become more important both for landowners and for all citizens.

Attempting to mandate good land management with a regulatory approach is certainly not the answer. Educating landowners about sustainable management and simultaneously educating urban citizens of the importance of well-managed privately owned rural lands are vital—and we must remain vigilant in both.

Social Stigma

There is as yet no social stigma in the possession of a gullied farm, a wrecked forest, or a polluted stream, provided the dividends suffice to send the youngsters to college.[5]

Aldo Leopold, 1938

In the years since Leopold wrote, there has been a great deal of progress in how we treat the land. Leopold lamented that in his day there was little or no sense of shame in the abuse of land. For most farmers and ranchers of the 1930s, poor management was so much the norm that it was not frowned upon. Damaged farms, ranches, forests, and rivers were accepted as the normal result of agricultural production. This was the era of exploitation. For those who did have a twinge of shame, poor land practice was easily justified by economic necessity.

What was not understood in those days was that it costs far more to restore eroded land than it does to control erosion in the first place. It is far easier and less expensive to apply basic conservation than it is to fix a wrecked landscape. However, our grandfathers and great grandfathers did not know this, and perhaps some did not care. The science of conservation was in its infancy, and the ethics of stewardship had barely been conceived.

Much has changed since those days. Today, as a result of widespread conservation efforts by private landowners, grasslands are in better condition, farms are more productive with far less erosion, forests are more stable and vigorous, creeks and rivers are healthier, and wildlife is more abundant and diverse. In Texas we see an increasing ethic of responsible land stewardship. A growing number of landowners consider it their personal obligation to care for the land with responsible and sustainable management.

However, there is still today some uncomfortable truth in Leopold's words about gullied farms and polluted streams. In some places there still appears to be little shame or disgrace when land is mistreated. We still see plenty of cases that should make us cringe. Overgrazing of ranges is still a problem in many places. The gross overpopulation of confined exotics is still taking place. Browse lines six feet high, pastures grazed as closely as a golf course, bare ground, eroding creek banks, unethical hunting, and other forms of land abuse are still with us and not easy to overlook or excuse. Nor should we attempt to.

We do not advocate a heavy-handed, forced approach to solving these problems. Instead, education, incentive, and positive peer pressure are better methods to bring about real and lasting change.

Jake Landers, well-respected rancher and conservation educator from Menard, Texas, has long advocated for a "kick-off award" to be given by conservation organizations. A gate sign would tell passersby that the offending landowners ought to be kicked off their land for abusive grazing practices. Landers insists there can be no excuse for extreme overgrazing and that private property rights should not include the right to abuse land. Perhaps his

Severe erosion on cropland still happens today but is much less frequent than in Leopold's day. Such natural resource abuse is increasingly socially unacceptable.

suggestion was only tongue in cheek, and public humiliation is probably not the best solution, but the idea of social pressure has merit.

In some farming communities, it is frowned upon to be a sloppy farmer with weedy fields and poor farming methods. These close-knit communities seem to have an unwritten code of ethics, and they take collective pride in good farming practice that is passed down to each new generation. Likewise in some wildlife management associations there is this same mutual expectation of good management and some degree of social stigma for poor management.

There is emerging a breed of landowner who is alarmed at land abuse by neighbors. These landowners, who practice good management on their own places, are concerned when they see blatant cases of mismanagement. After all, the reputation of all landowners suffers due to the poor practice of a few. Some of these individuals are starting to speak up. This form of peer pressure shows that our sense of collective land conscience is growing and deepening.

Most land in Texas today falls in between the extremes of abuse and excellent stewardship. The good news is that for every clear-cut case of land abuse we can find multiple examples of good conservation. The trend is positive and encouraging, and we must continue working carefully, respectfully, and creatively until good land ethics and good land management is the accepted norm.

Code of Ethics

We seem ultimately always thrown back on individual ethics as the basis of land conservation. It is hard to make a man, by pressure of law or money, do a thing which does not spring naturally from his own personal sense of right and wrong.[6]

Aldo Leopold, 1937

Each person has a set of ethics that governs their behavior. In some people, ethics are well developed, while in others they are not. Ethics involve our individual or collective sense of right and wrong and are driven by our inner convictions. Our ethics are influenced by our faith, traditions, family, education, friendships, and the culture around us. Our standards of ethics may change over time and become more limiting or less limiting. We may have a strict set of ethics in some areas of life and be lax in other areas.

Many occupations have a written code of ethics that define the responsible conduct expected for that profession. This includes doctors, lawyers, teachers, engineers, auditors, journalists, builders, and dozens of others occupations. Violating the code of professional ethics can be a serious matter. Our ethics, whether personal or professional, determine our reputation in our communities, families, and businesses.

In the world of land management and hunting, the public may get the wrong impression that all landowners and hunters engage in the questionable practices of a few. Setting forth our ethical standards may help separate the bad behavior from the norm. A code of ethics can be written or unwritten, but there is value in the effort of defining our ethical standards and, where appropriate, communicating them to others.

Some ethical questions have to do more with personal convictions than an inherent sense of right and wrong. For example, hunters will often limit the type of weapon they use for reasons of sportsmanship. Some bobwhite hunters shoot only 28 gauge and some dove hunters only .410 as a more challenging way of hunting. Some deer hunters no longer find high-powered rifles with sophisticated optics a challenging or satisfying way to hunt and have turned to archery or muzzleloaders. Some fishers would never consider using a baited hook to catch a trout or bass, while others have no scruples about it.

Hunters and landowners must also come to terms with the degree of artificiality they are willing to embrace in the raising and harvesting of wild game. For some, baiting deer with corn is not ethical, but others have no qualms about it. The issue of fair chase is a deeper ethical consideration. Most will agree that shooting a deer, elk, or exotic (such as red deer or scimitar oryx) in a hundred-acre enclosure involves entirely different ethical standards than hunting the same animal on a ten-thousand-acre high-fenced ranch.

This summary of ethics by an unknown author but often misattributed to Leopold is helpful and succinct: "Ethical behavior is doing the right thing when no one else is watching—even when doing the wrong thing is legal." Not everything that is legal is ethical. It is legal to shoot a duck on the water, a covey of quail on the ground, or a dove perched in a tree, but most hunters find these things unethical. Our ethical standards as hunters or wildlife managers will usually be more restrictive than what is allowed by law.

Landowners face a long list of ethical questions in the way they manage land. Will you plant nonnative forage grasses that are known to spread and crowd out native species? Will you aggressively deal with active erosion or bare ground? Will you give special protection to special places such as heron rookeries or turkey roosts? Do you allow hunters to place their blinds or feeders near the fence shared with your neighbor? Are you willing to diminish the flow of a neighbor's well by heavy pumping of water under your land?

Most landowners, hunters, or outfitters do not have a written code of ethics, but perhaps it would be a good idea. If we have lax or unspecified ethics and allow questionable practices, we are more likely to find ourselves and our activities scrutinized, criticized, and perhaps even regulated. Conversely, if we set high ethical standards for ourselves and stick to them, we will earn greater respect and favor as responsible stewards of the public trust.

Integrity, Stability, and Beauty

A thing is right when it tends to preserve the integrity, stability, and beauty of the biotic community. It is wrong when it tends otherwise.[7]

Aldo Leopold, 1947

Ethics are about the distinction between right and wrong, and Leopold declared emphatically what he thought was right and wrong in the biological world. The epigraph above is one of Leopold's most quoted passages and has become a motto of many ecologists and environmentalists. At first, it seems like a statement that most people could agree with in general. After all, who could be opposed to the preservation of biological integrity, stability and beauty?

However, upon further thought, the pronouncement may come across as too rigid and narrow regarding absolute right and wrong. Most things in nature and land management are not black-and-white. We usually work with shades of gray as we make land management decisions. We work with ecological principles that we only partially understand, not with cookbook recipes that describe the right way to manage land. The laws of nature are flexible because nature itself is dynamic. Let us consider Leopold's words in this light.

Integrity includes the quality of being complete or whole without any missing parts or defects and the keeping of those parts in working order. Most will agree that keeping all the parts of the soil-water-plant-animal complex is good. We have seen that the absence of such integrity can create problems.

But while absolute ecological integrity is a worthy goal, we must often be content with enough integrity to maintain the essential natural functions.

Most of the world's land has been impaired to some degree; it is not pristine. Yet, in most places, essential natural functions are still taking place while at the same time feeding, housing, and clothing the earth's population.

Stability normally indicates constancy, permanence, and steadiness. In nature, this kind of stability seldom occurs, but it is often expressed in terms of balance or equilibrium. The ecological quality of resilience seems to be more critical than perfect stability. Natural stability includes constant change in a way that keeps the system in relative balance, like two kids on a seesaw. If Leopold meant stability in the sense of maintaining a more or less balanced, resilient system, again most of us would concur that this is good, but absolute stability does not occur in nature.

Beauty is a relative and subjective quality, not a matter of right and wrong. The eye of the beholder defines beauty, and no one can define what constitutes beauty for another. What you may appreciate as lovely and beautiful may be very unattractive to others. Leopold's statement about the preservation of biological beauty comes across as too warm and fuzzy and is where he veers off track for some of us.

However, ecological beauty might be different from aesthetic beauty. Aesthetic beauty is determined by the eye and the emotions, while ecological beauty involves the mind and knowledge of ecological dynamics. Thus, a mature cedar brake mixed with oak on a steep hillside or canyon in the Hill Country is ecologically beautiful in the sense that this is what is natural and functional on that site. Removing the cedar and exposing the steep shallow soil to erosion and full sun would be unnatural, damaging, and therefore ugly from an ecological perspective.

Beauty can also mean a pleasant quality apart from visual appeal. Thus to a quail hunter, a scrubby patchwork of lotebush, mesquite, and sand shinnery, full of ragweed, croton, and sunflower, with a shaggy grass cover is beauty of the highest degree regardless of how "pretty" it is.

Likewise, few people buy postcards of spiny thickets of South Texas brush. To the untrained eye, it is hardly beautiful. Yet, to the person who takes the time to learn and understand South Texas, its beauty grows as the understanding deepens.

Leopold stated that biologic integrity, stability, and beauty should be preserved. Perhaps he only meant that natural processes should be maintained and not degraded. Today's land stewards generally prefer to use the term "conservation" rather than "preservation." Preservation implies little or no use or active management of natural resources. Conservation implies active manipulation and deriving beneficial and profitable uses within the context of sustainable management.

Aldo Leopold is rightfully revered and respected by conservationists across the globe. His writings are considered some of the most profound, prophetic, and meaningful in the world of natural resource management. However, we should be willing to detect the possible imperfections of what he wrote. Statements and opinions about what is ethically right and wrong are easy to make. But on close examination these rigid statements sometimes fall short. We should never make the mistake of putting an outstanding human on a pedestal no matter how inspiring one may be. We can and do admire Leopold even though we identify occasional flaws in his thinking. Doing this helps us think more clearly and accurately for ourselves.

When the Wells Went Dry

In the drought of the thirties, when the wells went dry, everybody learned that water, like roads and schools, is community property.[8]

Aldo Leopold, 1939

The drought of the 1930s is regarded as the worst large-scale natural disaster ever faced by the United States. There have been other catastrophic droughts since then but none of the magnitude of the thirties. Many lessons were taught by that drought.

Wells go dry for two primary reasons: extreme drought (which curtails aquifer recharge) and excessive pumping of groundwater. Either of these is bad enough, but when they occur together the results are especially adverse.

We now have many more wells than in Leopold's day, much greater water demand, and far greater pumping capacity, yet the total supply of water is the same. Simple arithmetic dictates that a finite supply of water cannot be infinitely pumped. When pumping exceeds recharge, bad things happen.

This lesson reinforces an important truth that landowners and society must embrace: *water is a shared resource.*

Community property is different than private property and harder to manage. When something is community property, it belongs to everyone yet to no one. There is a reduced sense of personal responsibility in caring for shared resources.

Interestingly, we have different laws that govern groundwater and surface water, even though the two are often hydrologically connected. In Texas, groundwater is owned by the landowner, while surface water is owned by the state. This makes it more challenging to manage and conserve our water supplies.

The ownership law in Texas is clear, and groundwater legally belongs to the landowner by the rule of capture, but many will agree that water, in a practical, ethical sense, is community property. We all benefit from it, and we all have a shared responsibility to use it carefully with regard to others and to the future. Groundwater is not confined by the property line and readily flows into and out from under tracts of land as a natural part of the water cycle. How your neighbors use their groundwater affects you, and vice versa.

Private property rights are an important part of our culture and legal system, and most of us will fight to protect these rights. But how can landowners protect their ownership of the water beneath their land? The truth is that you cannot protect your right of groundwater ownership unless the entire aquifer is being sustainably managed, and this requires the reasonable cooperation of all landowners for the common good. Otherwise the biggest pump wins, and your neighbors have the right to pump all the water they want, even if it causes aquifer levels to drop and your well to go dry. Reasonable, locally governed pumping limitations of privately owned groundwater are clearly beneficial and necessary where demand exceeds recharge. Landowners can and should become involved with their local groundwater conservation district to provide needed input into how shared aquifers can be best managed sustainably.

It all works fine when the rate of pumping does not exceed the rate of recharge on a sustained basis. Yet today some aquifers are in long-term decline, even with these regulations in place. It begs the question of whether greater restrictions are appropriate in order to sustain aquifers for the common good.

What happens in the skies often determines success on the ground. The uncertainty of rainfall and groundwater recharge means that we must use finite water resources with great care and restraint.

A false sense of water security was created in the 1970s, 1980s, and early 1990s when Texas experienced uncharacteristically favorable rainfall and aquifer recharge for a prolonged period. We were lulled into a sense of water abundance even though it was a climatic anomaly. Skipper Duncan, rancher, outfitter, and range philosopher of San Angelo, Texas, called these the "glory years of ranching" and admits that they created a faulty benchmark of what is normal.

Since that time, our rainfall patterns have reverted to a more realistic normal, including several very significant dry periods. Simultaneously more and more people are moving to Texas, and the trend is clear. The increasing water demand combined with uncertain rainfall and recharge is a calamity waiting to happen unless a deeper sense of conservation is instilled.

In the end, groundwater conservation is like other kinds of conservation. It is best accomplished by stewardship-minded landowners and water users who take their obligations seriously and use finite resources carefully and conservatively. Yes, groundwater ownership is a private property right in Texas, but we are also reminded of the other side of the coin. With every right comes corresponding responsibility. That is one of the primary facets of land ethics.

What Is Land Stewardship?

A land ethic . . . reflects the existence of an ecological conscience.[9]

Aldo Leopold, 1949

Conservation can accomplish its objectives only when it springs from an impelling conviction on the part of private landowners.[10]

Aldo Leopold, 1947

The subject of land stewardship has gained a great deal of positive attention in recent years, but, although "land stewardship" has come into popular usage, it is seldom clearly defined. When a term comes into such common use, there is a danger of it becoming an overused buzzword, obscuring and diluting its meaning. This lesson will offer a definition of land stewardship based on the ideals of Leopold as put into practice by many Texas landowners.

First, I should mention what land stewardship is not. Land stewardship is not merely the completion of practices, such as prescribed burning or native grass planting. These and other practices can be components of good land management, but the application of practices is not synonymous with stewardship. Land stewardship entails a person's relationship to the land. Stewardship is about who you are on the inside and what motivates your treatment of the land.

Leopold stated that true conservation springs from inner convictions about the land. He proposed that these inner convictions originate from an ecological conscience that defines what is right and wrong for each individual. Leopold taught that stewardship is driven by land ethics, and land ethics in turn drive our decisions and activities on the land.

I have had the privilege of working with many Texas landowners since 1976 and have learned a lot from them. While not all landowners demonstrate responsible land ethics, those that do are living examples of what genuine land stewardship looks like. They have taught me that the land steward is a responsible caretaker of the land. They consider themselves first and foremost to be custodians of the land rather than merely owners of that land. I have observed a strong sense of humility, respect, admiration, and love of the land expressed by these stewardship-minded landowners.

Based on these experiences with many Texas land stewards and the ideas of Leopold, land stewardship can be described as a deeply held inner conviction that compels and inspires people to be responsible caretakers of the land entrusted to them. The motivating forces for this stewardship are threefold: 1) present benefits to the landowner, including but not limited to economic benefits; 2) benefits to future generations; and 3) benefits that accrue to society outside the boundary of the land. Genuine land stewardship has a strong element of benevolence, and stewards realize that what they do on the land benefits others. These societal benefits may or may not bring financial return and may or may not be recognized by others.

Toward the end of his life, Leopold lamented that, as far as he could tell, no ethical obligations toward the land were being taught by schools, agricultural colleges, land bureaus, or extension services. This deficiency no doubt slowed

What happens on the private farms, ranches, and forestlands of Texas affects the vast Gulf Coast estuary habitats. In nature and in stewardship, everything is connected.

the development of stewardship ethics in the United States. Leopold's lament is still somewhat true today, but progress is being made, and there is good reason for optimism. Private nongovernmental organizations are leading the way, and some have established land stewardship as a centerpiece of their culture, purpose, and programs. It is significant to note that the stewardship message is being promoted primarily by landowners and landowner groups, not government conservation agencies. However, because of their passion and energy, the stewardship message is spreading to schools, universities, government agencies, and conservation organizations.

Everyone needs to understand the importance of land stewardship—city folks, country folks, children, adults, hunters, birders, fishers and anyone else who benefits from healthy land and clean water. Pitch in and volunteer when the opportunity arises. Seek opportunities to learn. There is no issue more important to the future of Texas than the conservation and stewardship of our lands, waters, and wildlife.

4

LEARNING THE LAND

Aldo Leopold knew that many good things can be learned from schooling and books. But he insisted that the land itself is the best textbook one could ever have and that with study and discipline one can develop an ability to read the land just as one learns to read a book. To be a student of the land requires keen skills of observation, the curiosity to look closely, the inquisitiveness to ask a thousand questions, and the desire to dig deep for answers. Leopold also taught that it is a virtue to profess ignorance when we do not know. This chapter contains ten lessons that teach the importance of studying the land to better understand how it works. As people gain understanding about the land, they will naturally be more motivated to care for it and apply thoughtful and effective management.

Healthy Curiosity

A good healthy curiosity is better equipment with which to venture forth than any amount of education.[1]

Aldo Leopold, 1920

Healthy curiosity: this is excellent advice to students, landowners, nature lovers, hunters, and biologists of all ages. Leopold states that a sense of curiosity about nature and land is superior to formal education and is the best way to continue and cultivate a lifelong habit of learning.

It is much better to be asking fresh new questions than it is to simply recite and recycle old answers. In natural resource management, we never gain a complete understanding of all the right answers. In fact, we often discover that the right answers are elusive and change over time. Some of the things we think we know today will be challenged and overturned in the decades ahead. Only by questioning our beliefs and opinions and practices with fresh curiosity will we move toward a better understanding of the complexities of the land.

In our world of specialists and experts, we often find too many canned answers and rigid formulas for land management. This is especially confusing for new landowners who don't know which experts to listen to. The know-it-all expert can be dangerous, while the best experts quickly admit that they have more questions than answers.

Humility and curiosity often go together. When one acknowledges that they do not have all the answers, it creates a curiosity and drive to seek better understanding. Inquisitive students may be closer to the truth than experts who think they already have the right answers.

Education is important, but it is not the most important thing. We all know highly educated people who can't get along with people, can't relate to the real world, and have no practical experience. Education plus humility and curiosity is a great combination for those who want to succeed at the highest level.

Leopold frequently mentioned to his students the importance of learning to read the land. It is a skill that requires curiosity, and a good teacher will always try to instill curiosity in one's students. Reading the land involves curiosity about the interrelationships of soil, water, plants, animals, and our management of these. It is a lifelong pursuit.

Reading the land begins with curiosity and requires a keen sense of observation, noticing things that may be hidden to the casual observer. This kind of observation does not happen very well from the front seat of a pickup. Walking the pasture slowly and pausing frequently to take a long close look is the best method for learning how to read the land. A slow, inquisitive half-mile walk can be more instructive than a full-day seminar.

Janna Blanchard and Bert Dieringer represent a large and growing segment of Texas landowners. They own a modest size tract in Mason County called Sideoats Ranch, and, while they are not ranchers in the traditional sense, they are true conservationists and serious land stewards. After successful careers, they wanted to spend their golden years helping others and caring for the land. They have availed themselves to countless educational programs teaching good land management, with good results. But the best education they have acquired is through their own personal curiosity and discovery. Janna, with insatiable curiosity, is forever searching for and finding new plants, which then leads to the desire to learn about a plant's niche and function and how it responds to differing conditions. Their curiosity about the land and how to manage it leads to many questions and the quest to find answers. In their stewardship journey they have learned one of the fundamental truths about land management—that there are always more questions than answers and that the process of discovery is the best way to learn the land.

While walking and looking, questions will naturally arise to the curious observer. Why is one area covered by threeawn while the area next to it is covered in blue grama? Why does one pricklypear plant look sickly while the next one looks vigorous and strong? Why do I see quail in this pasture but not in the next? Why do I see little bluestem growing adjacent to post oak but not out in the open? Why has this bare area never healed? Why do turkeys no longer roost in this grove of pecans? Why does the next ranch have pronghorn but we rarely see them on our place?

Curious land managers will have countless questions and will think about those things frequently, forming theories about what they see and why. Over time these theories will be tested, revised, refined, and often rejected. Little by little, the curiosity, the questions, and the observations will begin to provide insights and enlightenment. This is the process of acquiring land wisdom, an important element of good land management and an essential element of land stewardship.

Leopold and others have noted that most people are reluctant to venture very far off the beaten path. Let's resolve to spend quality time in the woods, hills, or pastures on a regular basis with a full sense of curiosity as we learn to read, interpret, and better understand the land we love.

Teach the Student

... Teach the student to see the land, understand what he sees, and enjoy what he understands.[2]

Aldo Leopold, 1942

Leopold was a gifted teacher, whether it was a wildlife management class at the University of Wisconsin, family and friends on a hunting trip, or a walk in the woods. Gifted teachers have the ability to instill enthusiasm and interest in their students and help them to see what others cannot.

Leopold was also a student. He was always observing, always curious, always thinking about the land, and never satisfied with his level of understanding. When it comes to land management, we are all students, no matter how well-educated or how many years of practical experience. We must never stop learning, questioning, and seeking a better understanding. This kind of diligent learning is not drudgery or mere classroom lecture. It is done with the enjoyment and enthusiasm that brings deep satisfaction as well as understanding.

Dalton Merz operates his family's small livestock operation near Holland, Texas. Merz has spent his career teaching landowners the principles and practice of sound range management, and his best teaching is conducted out in the pasture with the show-and-tell method. While he is teaching and demonstrating practical ecological truths, there is obvious excitement and passion that are contagious and make the student want to soak it up. The big smile on his face tells everyone that he loves what he does and loves to see others become excited about good land management.

Most of us are impressed by the intellect and ability required in highly skilled professions such as eye surgery, veterinary practice, and engineering. The profession of ecology and land management is just as complex and requires equal skill and intellect. It also requires constant and continuing education.

Both the student of the land and the teacher should be humble and honest enough to realize that the land is far too complex for anyone to completely comprehend. We know bits and pieces, and we understand enough to apply sound management, but we always strive to see the bigger picture more clearly and refine our understanding. Those who think they already possess all the knowledge they need are simply professing their ignorance.

An anonymous but astute ecologist once noted, "Those who do not understand nature are destined to deplete it." It is a sad reality, and we can see the truth of this proverb in our own state. As we drive along country roads we can observe that some landowners are still abusing and depleting their land. It is depressing as well as destructive. Such abuse of the land is not usually intentional; it is the natural outcome where there is a poor understanding of the land.

However, there is a contrasting truth that brings great encouragement: as we seek to understand nature better, we are compelled to conserve it. The more we understand about the land, the more we inevitably desire to take care of it.

As we consider our relationship to the land and to other people, we acknowledge that each of us is both a student and a teacher. There is always more that we must be learning, and as we learn we should teach others. Some of this learning and teaching is the science of the land, and some is the subtle art of management. In order for these to be effective, both the science and the art of conservation must be guided by a solid foundation of land stewardship ethics. Without the underpinning of ethics, science is merely head knowledge and art is merely for the emotions.

Leopold was both a professor and a poet. He poured his intellect as well as his emotions into his work of teaching and learning, and this is a good combination for today's land managers. It requires an investment of logical thought and scientific knowledge but also an investment of love, respect, and admiration for the land. Either of these without the other result in an unbalanced portfolio of management. Science and art work effectively in tandem but poorly by themselves.

Let each of us strive to see the land better by astute observation and to understand the land better by diligent study and inquiry. As we learn, we must also teach others. As we do these things there will naturally arise a greater enjoyment of the land, which in turn will motivate us to even greater love and understanding. The greatest enjoyment occurs as a result of a greater understanding.

Every Farm Is a Textbook

Every farm is a textbook . . . woodsmanship is the translation of the book.[3]

Aldo Leopold, 1943

The best wisdom is expressed in crisp, succinct phrases, not paragraphs or pages. Leopold was a master at this and was often able to boil down basic truths into a few concise words, as when he states that a farm is like a textbook. Think of an graduate-level textbook on a complex subject. Just owning the book or opening the book does not mean that one understands the material. It takes reading, rereading, digging in, studying, chewing, ruminating, digesting, and assimilating the knowledge it contains. Even after all of this, the student's understanding will still be incomplete until the information is put to practical use.

Like a textbook, a farm also contains a great storehouse of raw information, but learning to decipher and understand the information is not easy. It takes serious study over a lifetime, and new chapters of the farm textbook continue to be written year after year as conditions change.

The farms of Leopold's day were different from our present-day concept of a farm. Those farms were often a mix of row crops, forage crops, fruit and nut trees, large gardens, poultry, pigs, and other livestock. Farms often included woodlands, grasslands, wetlands, and creek bottoms full of many kinds of wildlife. The successful farmer had to be skilled in many types of agriculture and land management.

In reality, the farm textbook is more complex than the most advanced scientific book ever written. The farm is a text on soils, hydrology, watersheds, botany, agronomy, zoology, animal husbandry, range management, wildlife management, forestry, economics, marketing, social science, anthropology, and many other subjects.

Leopold states that "woodsmanship" is how the textbook of the farm is translated. This starts with being observant, paying close attention to everything and trying to figure out how it is all connected.

The word "woodsmanship" is most often used in the context of hunting, where it includes the ability to stalk, track, interpret signs, and understand animal habits and habitats. But in Leopold's usage of the term, it means more than just being a skilled hunter. In the context of the farm, it means the whole gamut of practical agricultural skills and abilities as well as the capacity to see the big picture of the farm, how it all fits together, and how it fits with the larger landscape outside the farm.

Leopold's "woodsmanship" means knowing the soil by observing the plant life. It means knowing what cows are grazing, what deer are browsing, and what quail are eating at each season. It means knowing the wild and domestic species that live there, including game species, rodents and other small mammals, birds, predators, insects, crops, livestock, trees, brush, grasses, and wildflowers. It means a basic knowledge of who eats whom and the sober realization that there must be death in order for there to be life.

Leopold was an admirer of the famous prairie ecologist John E. Weaver, who spoke of the ability to read the land. Weaver wrote, "Nature is an open book for those who care to read. Each grass-covered hillside is a page on which is written the history of the past, the conditions of the present and the predictions of the future."[4] Learning to read the land and comprehend the message is key to woodsmanship.

Woodsmanship is a close familiarity with the land learned by decades of curiosity, exploration, and getting one's hands dirty. It includes joys, disappointments, failures, successes, frustration, and satisfaction all wrapped together.

Some landowners never attempt to read or understand the land. They enjoy it superficially or aesthetically, and they may proclaim a love of the land, but the deeper lessons of the land may remain shrouded and undiscovered. Some miss out on the greatest joys of owning land. Let us make it our ambition to become more serious students of the land.

No matter what our role—landowner, manager, cowboy, biologist, hunter, fisher, birder, or nature enthusiast—we can dig deeper than before and hone our woodsmanship skills and begin to learn new lessons from the land. Let us slow down, look around, ask questions, seek answers, and approach the land with newfound hunger, thirst, and appreciation.

Understanding the Language

[Each species] has its own drama. The stage is the farm. The farmer walks among the players in all his daily tasks, but he seldom sees the drama, because he does not understand their language. Neither do I, save for a few lines here and there... . There is . . . drama in every bush, if you can see it. When enough men know this, we need fear no indifference to the welfare of bushes, or birds, or soil, or trees.[5]

Aldo Leopold, 1939

One of the necessary elements for a successful conservationist is a keen sense of curiosity, observation, and study. Education by attentive observation is far more interesting and of greater value than education by lectures, books, and seminars.

The language of the land is written primarily in plants. Dr. Dale Rollins, with Texas A&M University and the Rolling Plains Quail Research Foundation, frequently asks landowners how "fluent" they are in their knowledge of plants and their ability to read the land. The sobering truth is that many landowners, ranch managers, and biologists know a great deal about the animals they raise but comparatively little about the plants that sustain them.

Whether you are a landowner, land manager, hunter, or birder, your ability to interpret and understand the land is largely tied to how well you know plants. As you gain a greater understanding of plants, you will develop deeper comprehension of soils, wildlife, livestock, and the whole tangled web of ecology.

Leopold pointed out that the farmers of his day generally lacked much botanical knowledge and had not yet learned to study the land. Most farmers and ranchers of that era were engaged in the everyday routine of raising crops and livestock and trying to make a living in hard times. Ecological curiosity and enlightenment were not a priority.

Even today most landowners know five or ten brush species and maybe a similar number of weeds and grasses. Much of our plant knowledge is focused on learning how to control a few problem species rather than trying to understand and manage the hundred beneficial species.

Too often we delegate the need for plant knowledge to the professional conservationist working for an agency or organization. But Leopold insisted that the private landowner is in the position to be the best conservationist of a tract of land. Agency employees and professional conservationists have a role to assist and educate, but the leading role in conservation is always the landowner-caretaker who must decide what is best for one's place and then must live with the decisions made.

Neither a university professor such as Leopold nor the professional conservation agent can see or understand the full drama of a farm or ranch. They may see bits and pieces and may be able to enlighten the landowner for greater awareness, but it is the landowner who must figure out the drama on one's own place. As the drama is uncovered, the best direction of management becomes clearer.

Leopold discovered the drama of some of the plants and animals on his small weekend farm in Wisconsin, but he admitted that he had only a limited understanding. He did not live on the farm or make his living on the land. He knew that it requires years of careful daily observation to begin to unravel the interwoven drama on a piece of land.

The lesson is clear. The more we look, the more we will see and the better we will understand. For most people, a deeper understanding of their land will result in better stewardship. Take time to walk the pasture, deer lease, woods, or creek bottom, and do it frequently and in all seasons. Take time to notice and study things that are not readily apparent. Ask yourself questions and then seek the answers. You are the best ecological detective your land will ever have—if you are curious, observant, and diligent in learning the language.

Leopold concluded by saying that when people have the interest and make the effort to observe and learn the language of the land, the result will be a greater degree of careful management. Conservation will not just be the topic of speeches and seminars; it will be the lifestyle and lifetime pursuit.

Learning the plant life is a first step in understanding the land. On Leopold's visit to the King Ranch in 1948 he learned about the vast seacoast bluestem prairies and was impressed by the abundance of wildlife and productive grazing land managed in tandem.

The Personality of Silphium

The erasure of Silphium from western Dane County is no cause for grief if one only knows it as a name in a botany book. Silphium first became a personality to me when I tried to dig one up to move to my farm. It was like digging an oak sapling. After half an hour of hot grimy labor the root was still enlarging, like a great vertical sweet-potato. As far as I know, that Silphium root went clear through to bedrock. I got no Silphium, but I learned by what elaborate underground stratagems it contrives to weather the prairie drouths. . . .

   Why does Silphium disappear from grazed areas? I once saw a farmer turn his cows into virgin prairie meadow previously used only sporadically for mowing wild hay. The cows cropped the Silphium to the ground before any other plant was visibly eaten at all. One can imagine that the buffalo once had the same preference for Silphium, but he brooked no fences to confine his nibblings all summer long to one meadow. In short, the buffalo's pasturing was discontinuous, and therefore tolerable to Silphium.[6]

Aldo Leopold, 1949

Leopold spoke of plants as having personality, noting that each species has unique characteristics, preferences, and behavior. Leopold was a very capable botanist, and botany was the basis of much of his work in wildlife management, forestry, range management, and ecology. Leopold had the ability to read the land in the same way people read the printed word. The fundamental language of the land is plant life. When one learns to read the story written on the land by the plants, he or she is far along the path of land literacy. It is a skill that should be developed by each of us.

Silphium was one of Leopold's favorite plants. People of the land have favorite plants, just as they have favorite friends, favorite foods, favorite places, and favorite birds. Silphium is a perennial broadleaf forb in the sun-flower family, also called compass plant or rosinweed. Several species of silphium grow in Texas and exhibit the same characteristics as noted by Leopold.

One of the special characteristics of silphium, like many other deep-root-ed perennial forbs, is its extreme tenacity and durability. Leopold noted the ability of silphium to weather prairie drought. Many Texas forbs also have a seemingly supernatural ability to tolerate even the worst drought. Their roots penetrate deeply, not only to bedrock but even into and beneath the bedrock as their roots find every crack. The enormous taproots of these impressive prairie forbs store carbohydrates, enabling them to survive long periods of unfavorable conditions. Following the severe and devastating drought of 2011, when many grasses suffered or died, the deep-rooted perennial forbs fared very well. Texas forbs such as bundleflower, scurf pea, prairie clover, snout-bean, bush sunflower, trailing ratany, Engelmann daisy, gayfeather, gaura, and bloodberry, are able to withstand even the harshest drought.

Leopold also noted that cattle find silphium extremely palatable. In Texas, just as in Leopold's Wisconsin, silphium is most often noted on roadsides, where it has been protected from too-frequent grazing. Leopold took the time to watch what the cattle were eating and the context of the grazing. He noted that the prairie was only mowed occasionally and grazed infrequently, therefore silphium thrived. Even in what he called a "virgin prairie" full of the best grasses, the cattle grazed silphium in preference to everything else. The nutritional quality of many prairie forbs is often much superior to any of the grasses, and cattle as well as other herbivores seek them out. Many of these perennial forbs can be described as having a root like mesquite but with the nutritional forage quality of alfalfa. Silphium and all of the better grazing plants cannot tolerate continuous grazing, even at a low stocking rate. They thrive when grazed periodically and then given plenty of time to rest, recover, and regrow. This is the ecological basis for rotational grazing, which favors the better plants.

Many landowners in Texas are experts at observing and evaluating the condition of their livestock. A good rancher can tell in a moment the condition of individual cattle or sheep and whether they need any special attention. The astute deer manager can discern the relative age, health, and antler dimensions of a buck in only a fleeting glance. Good land managers also learn to develop the same sense of observation and discernment about the plants on their land. Out on the land is where such literacy is learned.

Emry Birdwell of the Birdwell Clark Ranch near Henrietta, Texas, is an astute observer of both animals and plants. He takes the time to intently watch what his cattle are grazing at all times, and he often records these observations with video. Even with a pasture full of fresh grass, his steers are sometimes seen to eagerly seek certain weeds and forbs such as western ragweed and prairie parsley, which are usually not highly regarded grazing plants—unless you watch closely.

Every plant on every acre has a personality, some delightful, some obnoxious, and some a mix of both. Some plants we wish to have more of and some less. Good land managers are those who know the plants on their place—their peculiarities, values, benefits, and detriments. They learn mostly by their own observations how the plants respond to management and what makes them thrive or decline. By studying plant personalities as well as animal habits and behavior, the manager begins to learn the subtle art of land management.

Pure Fire of Intellect

Conservation . . . is keeping the resource in working order. . . . In a surprising
number of men there burns a curiosity about machines and loving care in their
construction, maintenance, and use. This bent for mechanisms, even though
clothed in greasy overalls, is often the pure fire of intellect. It is the earmark of our
times.

  Everyone knows this, but what few realize is that an equal bent for the
mechanisms of nature is a possible earmark of some future generation.[7]

Aldo Leopold, 1939

Leopold frequently compared the land with the workings of a machine,
especially an internal combustion engine. In his day there was a select group
of men in every community who had an intense curiosity and special aptitude
to understand and tinker with engines. These were not the mechanical
engineers who designed and built engines but rather those who kept engines
running smoothly, in good working order. He said that these individuals had
a "loving care" for these machines and how they worked.

  He described this affinity and proficiency for mechanical things as the
"pure fire of intellect" of that era. We do not normally think of the mechan-
ically gifted guys in greasy overalls as being full of intellectual brilliance,
but Leopold did. The world of pistons, crankshafts, bearings, valves, lifters,
pushrods, rockers, cams, and gears coordinated with fuel, air, and spark is
a useful analogy to help us understand the complexities and inner workings
of the land. And the gifted mechanic is a good analogy for the skilled land
manager.

  To understand an engine you must take it apart and be able to put it back
together. You must know each part, its function, and how the parts all fit
together to run as a unit. Looking at a schematic diagram is not enough; you
must get your hands dirty. You must recognize when an anomaly or wear
affects the operation. Perhaps it is an almost imperceptible vibration, sound,
or hesitation that others do not notice, something you are able to trace back
to the source. The engine must be properly lubricated, coordinated, timed,

adjusted, and synchronized, and all of this takes human skill and intervention. It does not happen by itself.

The mechanisms of the land are also complex, even more complex than an engine. Every part is somehow tied to the other parts, but those links and relationships are often invisible. Nevertheless, the pieces must mesh and coordinate for balanced, productive, and sustainable operation. The land mechanic understands the practical connections between soil, water, plants, and animals and how these are affected, maintained, or enhanced by management. He or she is like the guy in overalls whose hands stay perpetually dirty with the everyday work of farming, ranching, forestry, range management, and habitat management.

Modern-day land mechanics are able to discern when something is out of sync and trace it back to its cause and prescribe a remedy through management. They use their skills, senses, experience, and intuition to determine when something is misaligned and how it might be corrected.

Perhaps quail habitat lacks adequate nesting cover, or maybe there is too much grass that smothers the forbs. Perhaps there are areas of bare ground that have never healed. Maybe the brush is starting to get so thick that pronghorn cannot adequately detect approaching coyotes. Meadow dropseed is increasing, and little bluestem is decreasing. KR bluestem is starting to encroach from the neighbor to the south. Spanish oak and black cherry are declining while persimmon and algerita are increasing. These and a thousand other irregularities are what land mechanics attempt to correct with management.

Leopold envisioned that in some future generation there might emerge a skilled group of people who have a special knack for understanding and managing land. In the years hence, his prediction has come to pass, and there are now scores of landowners across Texas who have that same aptitude and skill for land management as the exceptional mechanics of yesteryear. They have that same loving care, the same curiosity, and the same burning intellect that enables them to understand, manage, and sustain the complex natural ecological mechanism that we call the land.

The fate of this clutch of bobwhites is strongly influenced by the way the land is managed. The "land mechanic" is one who understands the inner workings of the land and who skillfully tweaks the mechanisms to produce the desired benefits.

The Basic Skill

The basic skill of the wildlife manager is to diagnose the landscape, to discern and predict trends in its biotic community, and to modify them where necessary in the interest of conservation. . . . To appraise the landscape the student must know its component parts and something of their interrelationships. That is to say, he must know its plants and animals, its soils and waters, and something of their interdependence, successions and competitions. He must know the industries dependent on that landscape, their effect upon it, and its effect upon them.

He must know and habitually use the visible "indicators" of those slow landscape changes that are invisible but nonetheless real. . . .

Last and most important, he should have developed in some degree that imponderable combination of curiosity, skepticism, and objectivity known as the "scientific attitude."[8]

Aldo Leopold, 1939

Leopold addressed the budding wildlife management profession in the 1930s, but his advice about knowing the land applies to landowners and managers regardless of whether they are raising crops, trees, or livestock, or fostering wildlife. Leopold said that the most basic skill is the ability to diagnose and treat the land. The similarity to the medical profession is clear.

The successful land manager must know about a lot more than just wildlife, livestock, or crops—he or she must know all the parts that make up the whole land system. When Leopold speaks of "the land," he always points out that this includes the soil, the water, the plants, and the animals as well as the interrelationships among these. Today's land managers must possess a set of skills that is wide and deep.

Not only must the managers know the parts of the land, they must also know how the parts fit together and function as a unit. They must know about the competition, the cooperation, and the synergism of the land. Just as medical doctors must understand the parts and systems of the human body, land managers must know the anatomy of the land as well as its physiology, chemistry, biology, and ecology. These are the skills necessary to make the land produce sustained yields of wildlife, livestock, crops, timber, water, and recreation.

Like medical doctors, land managers must be able to discern trends in the land, especially those gradual changes that are difficult for the untrained eye to see. They must be able to notice changes in plant abundance, plant vigor, plant diversity, and plant reproduction. Land managers must be able to relate these changes to the overall health of the land and to the crop or animal they are trying to produce or raise.

Not only should land managers be able to notice these changes, they must be able to understand what caused the change and what kinds of management and manipulation may be needed to correct any imbalances. This is where grazing management, brush control, prescribed burning, reseeding, and harvest management are applied. These tools and practices must be applied carefully with the right techniques, proper intensity, and correct timing. No two medical patients are alike, and two do not often respond the same way to treatment. Similarly, land does not necessarily respond the same way each time management is applied. Every situation is unique, and managers must apply uniquely customized treatments to achieve the desired results.

Chip Merrill reiterated this ever-changing facet of stewardship while he was managing his family's XXX Ranch near Crowley, Texas, saying, "What I am doing now, I wasn't doing last year, nor do I expect to be doing it exactly this same way next year." Ever-changing conditions require ever-changing management and constant examination.

Relatively few people have the opportunity and qualifications to be responsible, long-term land stewards. Only about 1 percent of Texans own land, and an even smaller fraction are engaged in genuine stewardship. The state's water supply, fish and wildlife resources, and crop, livestock, and timber production all depend on the faithful work of voluntary private land stewards. The work of successful land management requires a mixture of scientific knowledge, critical thinking, dedication, inspiration, and creativity. It is equally art and science and requires heavy investments of the mind, body, and emotions.

It is a high calling to faithfully carry out long-term stewardship responsibilities on a piece of land. For some, it is a life's work. The work is complex and challenging but also rewarding, personally satisfying, and beneficial to society. May each of us resolve to hone our skills, fortify our efforts, and focus our minds on the goal of taking good care of Texas.

A Serious Defect

The most serious defect . . . is the absence of the phrase 'we don't know.' Just why are we so undemocratic in the profession of ignorance?[9]

Aldo Leopold, 1937

Aldo Leopold was a teacher. Whether in the classroom or the field, he was always teaching. We think of Leopold primarily as a conservationist, hunter, ecologist, and philosopher, but he was also an educator and a very good one. He knew that he did not have all the answers.

Leopold's message about acknowledging ignorance was directed at the education profession of his day, but the point is applicable today to the fields of land and wildlife management. Most of us are reluctant to confess that we don't know. Those of us who make our living by providing conservation assistance and management recommendations are especially hesitant to admit we don't know.

Consider just a few of the things we still don't know when it comes to land and wildlife management:

- the long-term carrying capacity for deer on an individual piece of land
- the reasons lesser prairie chicken populations have declined
- the effect of juniper on the water cycle
- the role of soil fungi in regulating plant diversity, soil health, and productivity
- the reason deer in South Texas eat so much pricklypear compared to other regions
- the effect of fire ants on ground-nesting birds
- the accuracy of spotlight or helicopter surveys on individual ranches
- the reasons deer from granite soil areas are prone to hypogonadism and irregular antler growth
- the reasons some landowners possess a finely tuned land stewardship ethic, while others abuse the land.

We may have ideas and theories about these and a hundred other topics, but we have to admit that there is a lot we don't know. Mark Twain, among others, is purported to have said, "It's not the things we don't know that get us in trouble, but the things we know that just ain't so." In 1995 I confidently assured a rancher in Kimble County, Texas, that a cool-season prescribed burn would not harm his abundant mountain mahogany since it is known to be a prolific stump sprouter. I should have told him, "I don't know for sure whether fire will harm mountain mahogany." The fire was very hot, and it had not rained for several months before the fire, and it did not rain for months after the fire. As a result, nearly all of his prized mountain mahogany died.

Too often, we confidently proclaim the things we think we know, but often we are just repeating what we have heard or what we have been programmed to think. Sometimes we learn the hard way that we don't really know as much as we think we do.

The admission that we don't know is nothing to be ashamed of; just the opposite. When we admit we don't know something, we are more likely to continue searching and learning. Once we think we know the answer to a difficult or complex question, our minds become fixed and unable to see other perspectives. We lose the ability to say "I don't know."

We can all benefit from another old proverb of unknown origin: "He that knoweth not, and knoweth *not* that he knoweth not, is a fool—shun that man. But he that knoweth not, and *knoweth* that he knoweth not, is a wise man—follow him."

Humility involves the admission that there are many things we do not know. An attitude of humility on a regular basis will help us avoid giving wrong answers and foolish advice. It is perfectly fitting for us to honestly admit ignorance: "I don't know, but I will try to find out." It is the job of scientists to try to dig out the answers to the things we do not know, and scientists, too, are wise to exercise humility and continually rethink what they think they know. Oddly, it seems that the people with the least qualifications are often the boldest at proclaiming they have the answers.

There are many books that contain many truths that have been discovered. There are also many books, yet unwritten, that contain those things we do not know. We would do well to keep both kinds of books in our library of knowledge.

Dogmatic Statements

I have stated that any system of grazing, no matter how conservative, induces erosion. The proof of this statement . . . may be seen almost anywhere in the hills.[10]

Aldo Leopold, 1921

It is usually unwise to make bold, rigid statements about complex things. Natural resources and their management are unbelievably intricate, and new information is being discovered almost daily. There is much we are still learning even one hundred years after Leopold made his confident but incorrect statement that grazing causes land to erode.

We should temper everything we read or hear by considering the context in which it was delivered and with the realization that even the best scientists and land managers make errors and misunderstand things. A black-and-white statement will almost always prove errant, and some shade of gray will likely appear when a matter is more carefully examined and when the specific setting is considered.

The context of Leopold's statement about grazing causing erosion was the Prescott National Forest in Central Arizona, where rough topography, shallow soil, low rainfall, and limited natural water made it a challenging place to graze. In fairness to Leopold, what he saw at that time and place led him to conclude that any kind of grazing inevitably led to soil erosion. From where he stood, the statement seemed true. Everywhere he looked he could see signs of erosion on the hillsides that were grazed too short. But he made the proclamation as if it applied universally.

We should be slow to accept oversimplified statements about natural resource management, and we should be quick to consider other perspectives. We may hold on to our theories or the theories of others, but we should hold them loosely enough that we are able to let go of them when we see that they are flawed or incomplete. The most dangerous people in natural resource management are the ones who confidently assert that they have the right answers.

It is a mistake, of course, to automatically embrace what any person may say without question, but the same goes for experts. Most of us admire and respect Aldo Leopold for his insights and skills, but, like everyone, he sometimes made statements that later proved not to be completely true.

Early in his career, Leopold made dogmatic declarations in support of predator eradication. He aggressively promoted intensive control of coyotes, wolves, and mountain lions in the Southwest, knowing that it would mean more big game to hunt. What he failed to recognize at that time is that game populations unchecked by predators could—and would—destroy their own habitat.

He was young, enthusiastic, and overconfident. His enthusiasm was not yet backed up by the wisdom of experience or the ability to see the big picture.

Like many of us, as he matured he was less likely to make brash universal proclamations. With age we should learn to temper what we say.

We shouldn't be too hard on Leopold. Most of us have also embraced flawed thinking. In our day we have heard and probably adopted certain dogmatic ideas that later proved faulty. Here are some examples: Spikes and other "inferior bucks" should be culled to improve antler genetics. Since there is an 80 percent turnover of quail every year, hunters may as well harvest the "doomed surplus" since it will otherwise die of natural causes. Cedar and mesquite waste unusually large amounts of water, and the widespread control of these will dramatically improve our water supplies.

When we are tempted to make bold statements, we should be careful to qualify what we say and clarify the conditions under which we believe a statement is true. A little bit of humility goes a long way when we think we have some new revelation. It is wise to acknowledge "This is what seems to be true based on our current understanding." The test of time and further scrutiny will determine whether our beliefs stand, fall, or need to be modified.

With the complexities of natural resource management, it is frustrating to realize that the "right answers" sometimes change over time. But one thing that helps us navigate these complexities and frustrations is the resolve to stay focused on a strong and sustaining stewardship perspective.

Hundred Little Dramas

It is fortunate, perhaps, that no matter how intently one studies the hundred little dramas of the woods and meadows, one can never learn all the salient facts about any one of them.[11]

Aldo Leopold, 1941

Different forms of drama touch everyone's lives. Sometimes our world is disrupted by drama-filled people who drain our energy and create stress and disorder. There is drama at home, at work, at the hunting camp, and even in church. Sometimes we enjoy certain dramatic performances such as plays, television shows, and opera, but we realize they are usually only superficial, short-lived entertainment.

However, the kind of drama that Leopold describes is much different. The drama of nature is real and lasting; it enriches and teaches us as we study it. This ever-changing drama takes place continually in the pastures, hills, woods, marshes, and creek bottoms.

Leopold knew that people of the land will instinctively be engaged in observation and study, trying to understand the interwoven dramas of the land. Serious students of the land possess an insatiable inquisitiveness about the soil, water, plants, and animals and how these fit and work together under human management.

We have a thousand questions as we try to better understand the land. Why do my cows always do better in this pasture? Why do pronghorn only live on the north end of the ranch? How can I manage my grazing to grow more Indiangrass? What is the ecological role of desert termites? Is there a way to graze my pastures that will reduce KR bluestem and favor the natives? The longer we study the land, the more questions arise.

Bill Armstrong, the well-respected Texas Parks and Wildlife Department biologist who worked for many years at the Kerr Wildlife Management Area, used to say, "The book that contains everything we don't know is much thicker than the book that contains the things we know." This truth creates in us a sense of humility about our role as land managers as we admit how much we still have to learn.

Besides our quest for deeper scientific and practical knowledge of nature, there is another kind of drama that involves the emotions and keeps us connected to the land. The drama of the land is full of wonderment and awe if we take the time to seek it. Nature's drama can be exhilarating, calming, tumultuous, beautiful, and cruel, all in one day. Witnessing two big bucks fighting long and hard until one gives out, the other victorious. Hearing ten thousand sandhill cranes wake up on their roost, waiting for the sun to rise, and hoping that some will come your direction as you hunker down in a tumbleweed blind. Seeing the smile on a young boy's face after he shoots his first dove. Being suddenly surprised and revolted by a too-close encounter with a rattler, muttering words not fit to print. Watching a heifer give birth and

helping if needed. Smelling whitebrush, huisache or algerita in full bloom. Showing your grandchild how to bait a hook, then how to remove that hook from a wiggling sunfish. Watching the feeble and comical theatrics of a jake as he tries to court a hen. Going out the morning of the first hard freeze and seeing the exquisite ice sculptures formed on frostweed.

These and a thousand other little dramas fill our memories, inspiring greater awe and appreciation, making us want to know more, draw closer, and manage better. Nature is more miraculous and complex than we can fathom, but the more time we spend observing and studying, the more we

One of the unforgettable dramas of nature. The only thing that surpasses the sight of ten thousand sandhill cranes coming off the roost in the morning is their indescribable sound, primitive and eerie.

will comprehend. Little by little we gain some capacity to see the big picture of the land as well as some of the small details. The more we look, the more we will see. The more we learn, the more we want to learn.

What is even more humbling is to realize that we are part of nature's drama. We are not just spectators. Our attitudes and actions on the land determine if we fit into the workings of nature or if we upset the balance. May we ever strive to better understand, better appreciate, and better cooperate with the dramas of the land and teach others to do likewise.

5

LESSONS IN
CONSERVATION

The lessons in this chapter provide some of Leopold's diverse perspectives on conservation. As with any subject, examination from different angles gives a better picture than looking from only one position. Conservation today is more complex than is commonly understood, and Leopold provides an excellent grasp of its true nature including the challenges and difficulties. Included here are lessons relevant to the private land conservationist, conservation economics, the conservation agency, and public conservation policy.

A Positive Exercise of Skill

Conservation . . . is a positive exercise of skill and insight, not merely a negative exercise of abstinence.[1]

Aldo Leopold, 1939

The word "conservation" has been given various definitions over the years, and some of those meanings can cause confusion or misunderstanding. Leopold wrote a great deal about his own concept of conservation, and most of today's stewardship-minded landowners embrace his concepts.

Some of the modern viewpoints of conservation focus strongly on the idea of saving, preserving, and protecting. They tend to advocate a hands-off approach to land management. Lots of well-meaning nature lovers and environmentalists believe that conservation means little or no human intervention or use of natural resources. According to their philosophy, we should let nature take its course without interference or active management.

Leopold did not embrace this concept of preservation regarding management of private lands. He understood that the hands-off approach is not usually the best way to conserve wildlife, water, forests, grasslands, or any other assets of the land. The idea of a nature preserve or wildlife sanctuary without active management may sound appealing to one who does not understand natural resources, but it is short-sighted and idealistic. Taking active human management out of the equation is no different than removing other essential elements of nature.

Historically, there have been three primary ways that humans treat the land: exploitation, conservation, and preservation. Exploitation means that resources are used carelessly and indiscriminately with no sense of long-term sustainability. Natural resources are used until they are used up. Market hunting, extreme overgrazing, draining wetlands, and widespread deforestation are unfortunate examples of exploitation that were the norm of the past. We now regret these things and have learned from the mistakes of past generations. While there are still some cases of land abuse and exploitation, the trend toward better management is strongly positive.

At the other extreme, preservation means limited or no use of natural resources. Leopold called it the "negative exercise of abstinence." Abstinence may be a good solution for certain societal ills, but is not a good principle for land management. Preservation is a wasteful method of management that ignores the concept of renewable resources. Preservationists may have sincere intentions, but often they have a poor understanding of ecology, economics, and the needs of society. They want to save the environment without understanding the environment. They do not seem to realize the fact that humanity requires and nature provides for the consumption of natural resources.

A small East Texas sawmill where pine trees are converted to lumber. Leopold understood that land must be worked and its products harvested to provide necessary services and benefits to humans. When done correctly it is sustainable and perpetually renewable.

In contrast to exploitation and preservation, conservation strives to manage and use renewable resources in a way that maintains or restores these natural assets to meet human needs with an eye toward the future. Leopold said that conservation requires the exercise of skill and insight and that this necessarily includes an understanding of ecology, natural resource science, and management. Conservation is much more than just conducting recommended practices from a list. True conservation consists of doing the right things for the right purpose in the right places at the right time at the right intensity, with the commitment to ongoing management that will perpetuate and enhance the benefits. These things require not only our labor and money but also our skill, understanding, and a sense of conservation wisdom.

There will always be some tension when deciding the best form of natural resource management. Some will favor the more liberal use of resources to meet human needs, while others will favor the restricted use of resources in an effort to protect them from overuse. The conservation mindset is our best attempt to be responsible caretakers and active managers who use the renewable bounties of nature while preventing their degradation or depletion.

It is not an easy balance to maintain. With the guiding ethics of stewardship, many landowners are striving to find the sweet spot where resources are protected and enhanced while maintaining economic productivity and producing benefits for society. Texas is a better place when private landowners understand that they are the custodians and managers of the riches of nature—the soils, waters, plants and animals—and when they take that obligation seriously.

Intelligent Tinkering

Conservation is a state of harmony between men and land. . . . Harmony with the land is like harmony with a friend; you cannot cherish his right hand and cut off his left. That is to say, you cannot love game and hate predators; you cannot conserve the water and waste the ranges; you cannot build the forest and mine the farm. The land is one organism. Its parts, like your own parts, compete with each other and cooperate with each other. The competitions are as much a part of the inner workings as the cooperations. You can regulate them—cautiously—but not abolish them.

The outstanding scientific discovery of the twentieth century is not television, or radio, but rather the complexity of the land organism. Only those who know the most about it can appreciate how little is known about it. The last word in ignorance is the man who says of an animal or plant 'What good is it?' If the land as a whole is good, then every part is good, whether we understand it or not. If the biota, in the course of aeons, has built something we like but do not understand, then who but a fool would discard seemingly useless parts? To keep every cog and wheel is the first precaution of intelligent tinkering.[2]

Aldo Leopold, 1938

When Leopold spoke of conservation as harmony between people and land, he referred to a mutually beneficial working partnership—humans helping the land and the land helping humans. When Leopold spoke of the land, he meant the entire complex of soil, water, plants, and animals, and the inner workings of these. The harmony he referenced was not some fairy-tale view of nature but rather the relative sort of harmony that can also be discordant and harsh at times. This type of harmony appreciates the ugly cacophony in nature as well as the beautiful chords.

Leopold emphasized the complex inner workings of the land as vitally important to the proper functioning of the whole machine that we call nature. He was aware that people often label certain plants as weeds, certain shrubs as brush, and certain animals as vermin, pests, and "varmints." He taught that these less-popular native plants and animals are necessary, and without them the land organism is incomplete and dysfunctional. Mesquite, cedar, broomweed, coyotes, cowbirds, rattlesnakes, rodents, and desert termites are just a few of nature's essential parts that are often maligned and misunderstood.

Leopold also acknowledged that humans have been placed in the unique position to regulate the inner workings of the land. This regulation is what we call management—the manipulation of the system for a desired outcome. However, Leopold was quick to point out that this managerial regulation and manipulation of nature should be done cautiously and with much care and thoughtfulness.

Chad Druen-Miller manages several ranches in Central Texas for absentee owners. He understands the importance of being cautious and thoughtful since he is responsible and accountable for the care and condition of someone else's land and the wise expenditure of their money. He must be able to explain to the owners the logic and purpose of the management he applies: why he selectively thins the density of mesquite rather than controls it; why he judiciously sprays some pricklypear clumps while leaving others intact that are protecting desirable forbs; why he carefully chooses the location, size, and layout of food plots for maximum benefit to multiple species and to prevent erosion; why he carefully selects equipment operators and ranch laborers based on their skill and the quality of work they do. Intelligent, cautious, and thoughtful manipulations of the land are the earmark of successful stewardship.

This cautious and thoughtful regulation of natural processes is what Leopold described as "intelligent tinkering." Any youngster with the curiosity and inclination to take apart machines quickly learns the necessity of keeping all the parts. Discarding or losing a seemingly insignificant part has ruined many a good machine or at least hindered its useful operation.

The skillful mechanic not only keeps all the parts but also learns the name, function, and characteristics of each and how they fit together and cooperate. Wise land managers are like master mechanics or perhaps skilled surgeons. They study and understand the machine, be it an automobile or a human body, before they decide to fix or adjust it.

Yet the land organism is more complex than any machine and even more complex than the human body. We have the ability to fix most machines if they are broken; medical science has the ability to repair most injuries and treat many diseases. Those who wisely and properly regulate the mechanisms of the land to maintain harmony and sustain the balance are what we call conservationists. Conservation is as much an art as it is a science, and it requires as much skill, dedication, and wisdom as any profession on earth. Whether we own land or not, let us each strive to support, encourage, and promote intelligent tinkering on Texas lands for the benefit of all Texans, now and for generations to come.

Acts of Conservation

During the CCC epoch, many Wisconsin farmers were induced, by subsidy, to perform the acts of soil conservation; but those who lacked desire and skill dropped the acts as soon as the subsidy was withdrawn. . . . Acts of conservation without the requisite desires and skills are futile.[3]

Aldo Leopold, 1944

Conservation was a new concept in the 1930s and was not yet a normal part of agriculture or land management. The Civilian Conservation Corps (CCC) was one of the New Deal programs ushered in during the Franklin Roosevelt administration. It put people to work during the Great Depression and installed public infrastructure projects and conservation practices on private farms.

In order to attract attention and stimulate interest, subsidies and free labor were offered to get farmers to try conservation techniques with the hope that they would be adopted as an ongoing part of the farm operation. In the early days, subsidies were provided to build terraces, waterways, and diversions and check dams—and to plant trees and implement other conservation practices.

Since then, subsidized conservation programs have expanded greatly in number and scope. We now have many generously funded state and federal incentives for many different kinds of conservation on croplands, rangelands, forestlands, and wetlands.

But how many of these cost-sharing programs actually result in a long-lasting conservation mind-set and a continuation of the practices once the subsidy is gone? That was the question in Leopold's day, and it is still a valid question today.

Individual acts of conservation are good at least for short-term benefits and to jump-start the adoption of new ideas. These acts of conservation are comparatively easy, especially when subsidized. But it is much harder to adopt and maintain a coordinated system of conservation that involves the overall health and management of the soil, water, plants, and animals.

Leopold stated that individual acts of conservation are pointless if the owner lacks the inner desire to continue and unless one attains the necessary skills to do it on one's own. In many cases, even today, such programs entail impressive financial assistance but lack the more important management components that must go along with the practice in order to achieve the intended benefit.

Leopold wrote, "Subsidies and propaganda may evoke a farmer's acquiescence, but only enthusiasm and affection will evoke his skill."[4] A government incentive may be what sparks an interest and gets the ball rolling, but there must be something deeper. For true and lasting conservation success, ongoing practices must be motivated by an inner land ethic, not merely the financial assistance.

Conservation programs are funded by the government and justified to taxpayers on the basis that they provide direct or indirect benefits to society, not just payments to landowners. There must be the assurance that the practice will be maintained and that the landowner will develop the desire and skills to perpetuate and sustain what has been funded.

A complex alphabet soup of agencies, organizations, and programs encourages and assists landowners to carry out conservation. The programs are good when customized to individual landowner needs and when they fit in with an overall conservation strategy of the farm, ranch, or forest tract. Too often, though, agencies equate conservation success to program participation numbers and dollars spent.

For agencies, successful programs require capable, trustworthy, and committed employees at the field level and all administrative levels. Building lasting cooperative relationships with landowners is far more important than program participation. The relationship must be built on trust and technical competence, not just financial benefit. When agencies and their employees understand this, their programs and incentives can become an important conservation catalyst and a benefit to society.

Successful, enduring conservation is a way of life, not just enrollment in a program and installation of practices. With or without financial assistance, lasting conservation requires skill, dedication, sacrifice, risk, and investment on the part of the landowner.

Lopsided Conservation

Lop-sided conservation is encouraged by the fact that most Bureaus and Departments are charged with the custody of a single resource, rather than with the custody of the land as a whole. Even when their official titles denote a broader mandate, their actual interests and skills are commonly much narrower.[5]

Aldo Leopold, 1944

Narrow interests, what Leopold called "lop-sided conservation," caused him much frustration. The agencies that were charged with managing land or assisting landowners were guilty of tunnel vision and were unable or unwilling to see the big picture.

Long before the word "holistic" became popular, Leopold insisted that we must look at the land as a whole rather than just considering its individual components. This is true on farms and ranches just as it is true in the culture of government agencies. He lamented that the agencies of his day were guilty of promoting conservation based on their narrow mandates and abilities. The agencies were out of balance, and this contributed to lopsided management of farms, ranches, and forests.

Soil, water, plants, animals, and air do not exist in isolation from each other. They are interdependent and complementary and strongly influenced by human management. A true understanding of one component is impossible without a solid grasp of other associated parts. This is ecology—the study of all of the components of a natural system and especially the interrelationships and dynamics between the parts. It seems intuitive to us today, but it was not well understood in the mid-twentieth century.

Fortunately, much has changed since then. Perhaps the agencies heeded Leopold's reprimand. In most cases, their philosophies have evolved to incorporate a greater degree of holistic thinking.

We owe a debt of gratitude to Allan Savory who brought the concepts of holistic resource management to the United States from Zimbabwe. The holistic principles that he developed and that are still being refined are the foundation of what is now being called regenerative agriculture. Savory was and still is a controversial figure, being equal parts of brilliant and abrasive. Loved by some and criticized by others, he has propelled a paradigm shift in the way that natural resource management is viewed on millions of acres and within many agencies and organizations.

The Soil Conservation Service once focused exclusively on reducing soil erosion. It was and still is a worthy cause, but it was only one piece of the puzzle. Now known as Natural Resources Conservation Service, the agency has expanded and deepened from a single issue to a much better understanding of how the parts must all function in sync under the management of the private landowner.

What we now know as the Texas Parks and Wildlife Department (TPWD) started out as the Texas Game, Fish, and Oyster Commission. At that time they were charged primarily with establishing and enforcing hunting and fishing laws. Today TPWD is one of the most multifaceted and capable agencies in Texas with a wide range of natural resource expertise and ability.

The Texas Agricultural Extension Service was almost exclusively involved with agricultural production, 4-H, and improving farm life. In the early years it did not delve into conservation of natural resources or wildlife management. Now Texas AgriLife Extension is involved in a whole kaleidoscope of issues relevant to agriculture and natural resource management.

These agencies as well as many conservation organizations started out with a narrow field of vision that kept them from seeing the whole picture. The agencies and organizations sometimes even worked at cross purposes. To their credit, and to the credit of the governing bodies that direct and fund the agencies, they evolved and now embrace a greater degree of holistic thinking. In many cases it has been innovative landowners leading the way, urging the organizations to make needed changes. Regardless of the cause, there have been positive changes.

Whether you are a landowner, birder, hunter, business owner, or agency employee, you must strive to see the world through a wide-angle lens rather than a telephoto. Yes, it is good to study a discipline up close and in detail, but then you must back away and examine the bigger picture.

On a ranch, our focus must be much broader than just cattle production, quail populations, or big antlers. We must also consider the health of the soil, the integrity of the water cycle, plant diversity, economic sustainability, estate planning, personnel management, infrastructure, and cooperation with neighbors and the community.

It is easy to find fault with government agencies, with their bureaucracies and inefficiencies, but in the world of land and wildlife management we must commend the agencies that have stepped up to provide better and more balanced service. Texas lands and people are better off because of it.

The Heart of Conservation

The first thing to grasp is that government, no matter how good, can only do certain things. Government can't . . . bring to bear . . . that combination of solicitude, foresight, and skill which we call husbandry. Husbandry watches no clock, knows no season of cessation, and for the most part is paid for in love, not dollars. Husbandry of somebody else's land is a contradiction in terms. Husbandry is the heart of conservation.[6]

Aldo Leopold, 1942

Leopold reminded his readers that governmental conservation agencies and programs can be useful, but they can never compare to private conservation carried out by private landowners. During the New Deal era of big government, it was common for agencies and bureaus to take over some of the jobs that private enterprise and private landowners could have been doing without government assistance.

In the years since Leopold wrote, some things have not changed. Government agencies still have programs to assist, encourage, or subsidize conservation activities. While many of the programs are beneficial, they too often fall short because they do not touch on the long-term commitment of private husbandry that must accompany responsible and successful land management.

Leopold uses the word "husbandry" to describe the quality by which private landowners voluntarily take care of their land for the long term. Although we do not use the term much these days in this sense, husbandry represents the same idea that we now call stewardship. It is the inner conviction that inspires and motivates landowners to be responsible caretakers of the lands entrusted to them.

The vast private forestlands of East Texas provide wood, water, wildlife, recreation, and the storage of atmospheric carbon in the soil. A strong sense of husbandry motivates landowners to take care of these natural assets, sometimes with the help of conservation incentive programs.

Conservation does not occur just because landowners have enrolled in a conservation program that compensates them for implementing some conservation practice. Landowners are naturally glad to have financial assistance to help cover the costs of conservation work. Likewise, they are grateful to receive incentives to better manage their livestock grazing, soil health, or forestland, but, without the underlying ethics of husbandry, the benefits are likely to be superficial or short-lived. Unless there exists a deep internal sense of husbandry on the part of the individual landowner, government programs will never achieve the lasting benefits for which they were established.

As a thirty-five-year employee of a federal conservation agency, I observed the plain truth of Leopold's statements. Government agencies can do some helpful things but never at the same level or with the same effectiveness as what can be accomplished by conscientious landowners. Part of the reason for this is explained by Leopold—there can never be the same degree of attentiveness, insight, or enthusiasm from the conservation agency as there is from the landowner. "Husbandry of somebody else's land is a contradiction in terms."

There are many well-trained, dedicated employees of government conservation and wildlife agencies. Most of them do the best they can within the parameters of their agency, but rarely do you find a government employee exhibiting the same inner drive, work ethic, or love of the land as conservation-minded landowners.

There is something much different being a professional agency conservationist, specialist, or wildlife biologist than being the actual landowner. The agent can observe, suggest, and provide guidance and technical expertise. The agency can provide cost-share, grants, or other managerial incentives, but, without the daily involvement, interaction, and investment in the land, the most essential element is missing.

In the same essay as the source of the quotation, Leopold added, "When we lay conservation in the lap of the government, it will always do the things it can, even though they are not the things that most need doing." This is not so much a criticism of government but simply an expression of the obvious truth that government is unable to do the most important thing—husbandry.

In Texas we are fortunate that private landowner stewardship ethics are growing stronger. In my years of working with private landowners, I have seen a deepening of the husbandry ethos which drives true conservation. Government programs and priorities ebb and flow, their funding is erratic, and the ever-changing bureaucratic rules can be exasperating, yet these can serve a beneficial purpose when combined with the husbandry ideals of private landowners.

A Bird Dog Named Gus

I had a bird dog named Gus. When Gus couldn't find pheasants, he worked up an enthusiasm for Sora rails and meadowlarks. This whipped-up zeal for unsatisfactory substitutes masked his failure to find the real thing. It assuaged his inner frustration.

We conservationists are like that. . . . To assuage our inner frustration over this
failure, we have found us a meadowlark. . . . It smells like success. . . . There is
danger in the assuagement of honest frustration; it helps us forget we have not yet
found a pheasant.[7]

Aldo Leopold, 1938

Leopold's description of conservationists' frustration is one of his more
sobering passages. In our zeal to be perceived as good conservationists, we
sometimes substitute the trappings and illusions of conservation for the real
thing. His message is true today, just as it was in his day. Leopold was not just
pointing the finger at others—he included himself in the rebuke.

Landowners, their managers, and advisors sometimes confuse the implementation of certain practices with true conservation. There is a big differ-
ence. For example, brush or weed control does not necessarily result in conservation, and rotational grazing or fire might or might not deliver improved
grassland conditions. Any tool, practice, or method can be used wisely for
benefit or unwisely for harm. The tools and practices are neither good nor
bad—they can be either, depending on the level of skill and commitment with
which they are applied, maintained, and managed. These various practices
are not by themselves the criteria for conservation.

Wildlife managers are also guilty of succumbing to a shallow sense of
pseudo-conservation accomplishment. Placing a few birdhouses is a poor
substitute for having an abundance of natural nest cavities. Creating pollinator gardens is nowhere near as beneficial as having high native plant diversity
across the landscape. Feeding songbirds, quail, turkey, or deer has almost
nothing to do with conservation and sometimes does more harm than good.
But, like Gus pointing meadowlarks, these things make us feel good. They
are easy and give the appearance of success; we have done those things on
the conservation checklist.

Conservation agencies and organizations can also be guilty of this failure.
Naturally they are anxious to convince the public that their department has
done a good job, and they cite statistics to prove it. "Our employees wrote
five hundred wildlife management plans last year on two million acres."
"Our staff prepared twelve hundred cost share agreements and obligated

$20 million." "Our group conducted sixty prescribed burns on twenty-five thousand acres." "We worked with 150 new landowners and certified their property as wildlife-friendly."

It all sounds impressive, and no doubt some real and lasting good is accomplished through the help of organizations and their programs. But program participation and numerical reporting is not a good measure of conservation. A boy may wear a Boy Scout uniform, but if he does not embrace the character of scouting and cannot paddle a canoe in a straight line one must wonder if he is the real thing.

Let us strive to be real and honest in our recognition of conservation achievement. We should commend those who are getting it right, but we should not be too anxious to pat ourselves or others on the back. Real conservation is much more than going through the motions and will always elude those who try to make it too easy.

Real conservation is not easy to define and impossible to quantify. It is a lifestyle built on understanding and caring for a piece of land. It involves a long-lasting relationship with the land characterized by knowledge, skill, commitment, creativity, patience, perseverance, and adaptation.

Genuine conservation is alive and well in Texas, and it is growing. In our pursuits of conservation, let us chase after what is real and not be satisfied with easy, unsatisfactory substitutes. The results are worth it.

Uneasy Conservation

We lost the Flambeau as a logical consequence of the fallacy that conservation can be achieved easily. It cannot.[8]

Aldo Leopold, 1947

In an essay written near the end of his life, Leopold laments the loss of Wisconsin's Flambeau River when economic opportunity took precedence over the value of a natural, free-flowing river. Unrestrained logging and the rapid growth of the dairy industry had already taken a heavy toll on the once great forests along the Flambeau. Then in Leopold's day a series of hydroelectric dams threatened the character of the river itself. Leopold and his allies lost their battle to block the damming of the river's last remaining natural stretch. It was a bitter reminder that big wins for conservation are not easily achieved, and losses are often far-reaching and sometimes irreversible.

Leopold was not against logging, dairy farming, or rural electrification, but he urged people to consider that economic progress was not always the right move. Leopold knew that economics is important, but he argued that it should not be the sole factor in making natural resource decisions.

We face the same dilemmas today, not only for public projects but also in the legislative and bureaucratic arena where laws are enacted and conservation policy is decided. Good conservation and good economic opportunity sometimes fit together nicely—but sometimes they do not. Sometimes battles must be fought, and sometimes economic opportunity favoring one segment of society must be sacrificed for the greater, lasting good.

Sometimes the solution entails negotiated compromise, with both sides sacrificing some of what is important to them. Other times compromise is not acceptable. Some issues are worth fighting for and require rolling up the sleeves, doing the hard work of changing minds, and gaining both popular and political support.

The pressures we face today are substantially greater than those faced by Leopold. There are now more than twice as many people in the United States as in Leopold's day, and in Texas we now have four times more people than in 1947, yet the total land mass and water supply has not changed.

As our population grows and as cities and towns encroach into the countryside, there will be more new roads, pipelines, powerlines, energy development, water and sewage treatment plants, landfills, gravel mines, commercial development, and other kinds of infrastructure. These things take a toll ecologically, aesthetically, and culturally just as the hydroelectric dams did in Leopold's day. How much are we willing to lose? Is unrestrained economic development worth the cost it extracts on the natural world? These are questions that we must face as Texas conservationists.

Disagreements in natural resource management are inevitable. Conflicts between conservation and economics are not black-and-white and not easily reconciled. Real conservation solutions are seldom simple or straightforward and often require sacrifice and battles. Conservation does cost something, and it can be painful in the short term. But the conservationist takes the long, studied look and considers what is better in the long run.

What is better for Texas now and in the long run is abundant clean water used carefully and conservatively, intact functional watersheds, productive and sustainable agriculture, abundant and diverse native wildlife, and people who understand and appreciate nature and how nature works.

Life teaches us that nothing good comes easily. Work, sacrifice, and endurance are requirements of successful living. These qualities are nowhere more apparent than in agriculture, conservation, and wildlife management.

Providing food, fiber, water, wood, outdoor recreation, fish and wildlife habitat, and other necessities is more challenging than ever. Are these forms of natural infrastructure any less vital that urban and economic infrastructure? I think not. Let us muster all the strength, creativity, and determination we can to retain what is special about Texas.

An Intense Affection

Perhaps no one but a hunter can understand how intense an affection a boy can feel for a piece of marsh. . . . I came home one Christmas to find that land promoters, with the help of the Corps of Engineers, had dyked and drained my boyhood hunting grounds on the Mississippi river bottoms . . . My hometown thought the community enriched by this change. I thought it impoverished.[9]

Aldo Leopold, 1947

Young Leopold was heartbroken to find his boyhood duck hunting grounds lost to progress. Leopold's duck marsh was destroyed by diking and draining the lowlands along the Mississippi River near his hometown of Burlington, Iowa. In those days, wetlands were considered wasted land that could be put to better use with some engineering and earth moving.

Of course the value of a marsh is much greater than simply duck hunting, but don't try to explain that to a young duck hunter. Duck hunting, or any kind of hunting, gets in your blood in a way that is unexplainable. Nonhunters cannot understand it, and hunters cannot logically explain their affections for wild places.

Leopold's sad story can be repeated thousands of times in every corner of America. It is not just the loss of marshes and hunting grounds but also the loss of all natural landscapes and the benefits they bring.

My brother Doug and I hunted doves relentlessly in rural Denton County in the late 1960s and early 1970s. It is what we lived for and dreamed about. We had free run of several sections of old cropland, fencerows, woods, grassed fields, ponds, and creeks surrounding our small family tract. At that time this area of modern-day Flower Mound was rural and agricultural, but it was in transition and about to become urbanized. We had no idea how drastic, rapid, or depressing the changes would be. Each time we would come home from college, we would see new losses and more of our hunting grounds gone forever.

It was not just about the hunting. Here was where we explored, shot at tin cans, solidified friendships, deepened family bonds, and worked on the family property with Dad. It was where we learned what it means to have a relationship with the land. The memories of many successful hunting trips are still fresh, but, as any hunter will tell you, hunting is not all about the kill. What we remember most are the experiences, not the number of birds in the bag.

Some say it is progress. Some say it is the inevitable price we must pay for greater economic growth. But when your boyhood hunting grounds are swallowed up and destroyed, those explanations offer no consolation. Some insist that the needs and wishes of affluent society must take precedence over habitat for wildlife and other natural values. For those with no attachment to land, such losses are easy to justify. For those with deep attachments and personal history with the land, the losses are deep and painful and impossible to reconcile.

Many an aged farmer, rancher, or forest owner has shed a tear when family land is sold and lost to development. They feel the loss acutely. Perhaps they did not adequately instill a loving connection to the land in their children or grandchildren, or perhaps the attachment was not deep enough to offset the monetary allure of selling the land.

Where there is weak affection for the land, the loss of marshes, woods, grasslands, and farmlands is not considered that great a loss. But there is a cumulative and compounding effect of these losses piling up over time and across the landscape. It's not just the loss of hunting grounds; it's the loss of water-filtering and water-absorbing landscapes, the loss of aquifer recharge, the loss of native plant and animal diversity, the loss of carbon-absorbing lands, and the loss of agricultural production. And just as important are the lost opportunities for people to develop personal connections with the land.

One day society will discover that such natural losses are as real and as important as economic growth.

Internal Controversy

The love of nature is a matter of affection and esthetics. The understanding of nature is science. The use of natural resources is economics.

The conservation movement is composed of people in whom these three elements are mixed in widely varying proportions. It follows that there will be internal controversy. . . .

The controversial nature of this field sets up this specification: All three elements must be represented.[10]

Aldo Leopold, 1939

In describing the interplay of aesthetics, science, and economics when it comes to nature, Leopold demonstrated his understanding of the complexity and internal struggles in the discipline of conservation and land management. Our interest in land and wildlife management may start out in any of the three places, but as it matures it must involve all three elements—it must touch the heart, satisfy the mind, and consider the bank account. This balance should be sought in individuals, organizations, universities, and agencies, as well as farm and ranch enterprises. An overemphasis or under-emphasis on any one element results in a deficiency or excess in another element. It is a balancing act that is difficult to achieve and even harder to maintain.

An interest in conservation sometimes starts out with an appreciation of outward natural beauty such as flowers, birds, butterflies, rivers, and sunsets. These affections draw us to the land and provide intrinsic value. Admiration of outward beauty can be a good introduction into nature, but that alone will not take you very deep into the world of conservation. It may be a starting place, but it must go beyond the aesthetic.

Some people pursue nature as a scientific quest, striving to learn as much factual information as possible about the natural world. Fortunately, there are many great scientists in Texas who help us understand natural resources. Leopold himself was a brilliant scientist, using the science of ecology as the basis for his work in forestry, range management, watershed management, and wildlife management. Today wildlife ecology is more complex and involves the integration of many disciplines, including soil science, botany, animal science, hydrology, agriculture, and sociology. It requires a sharp mind and focused thought. But good science cannot stand alone. People may spend their entire careers in the sciences making important contributions, but if they lack a connection to the emotional elements of the land or economic realities their scientific contributions will be stunted.

For other people, the primary focus of land management is economic. They rightly emphasize that the land must generate enough income to cover costs and provide a return to the owner. This is especially true for landowners who make their living from the land. It costs money to own land, and it costs money to operate and manage land. Most landowners do not have enough outside income to completely subsidize their land ownership expenses, and therefore the land must either pay its way or generate enough revenue to justify ownership.

Alan Curry, who operates family ranches near San Angelo, Texas, explains the necessity of livestock-generated revenue that supports active and ongoing management and maintenance. He says that day-to-day livestock management is often what keeps people living on the ranch. If livestock are removed and people move off the ranch, watering systems will break down, fences will deteriorate or become overgrown, roads and water gaps will get washed out, and barns, pens, houses, and headquarters areas will fall into disrepair. He insists that landowners must have a way to generate sustainable income to stay on the land and pay for its ongoing management. Livestock are one very important way that enables land to remain profitable and well managed.

Scientific knowledge and emotional affection will not pay for the considerable expenses of conservation and land ownership. But economics alone without the ecological and aesthetic elements can be short-sighted and lead to exploitation and degradation.

When there is a love of nature and an understanding of nature and the economic use of natural resources, then a kind of harmony and sustainability becomes possible. An imbalance among these three factors creates stress and tension and hinders long-term conservation and productivity. The heart, the mind, and the pocketbook are all essential parts of a balanced conservation equation.

Leopold said that "conservationists are notorious for their dissensions."[11] He did not say that this dissension was bad. Dissension is a fact of life and helps us see other perspectives. Today there are plenty of disagreements in Texas land management, and sometimes the disagreements are heated. If we try to address our dissension with a proper combination of aesthetics, sound science, and economics, seeking what is good for the whole rather than the individual, resolution is possible.

Successful conservationists have a diverse portfolio of investments—they invest their love and affections, their scientific understanding, and their economic resources into the land. The result will be a beautiful, attractive landscape, functional ecosystems, and economic productivity, all of which help sustain the lands, waters, wildlife, and people of Texas, now and in the future.

The Economics of Conservation

It of course goes without saying that economic feasibility limits the tether of what can or cannot be done for land. It always has and always will.[12]

Aldo Leopold, 1949

Leopold wrote about hunting, game management, agriculture, songbirds, botany, conservation, ethics, education, and many other topics. He also wrote about economics, and he understood that economics dictate much of what can be done on the land. He knew that money does not grow on trees and that the land must generate income in order to pay for the costs of owning, maintaining, and managing the land.

Too often, real-world economics is divorced from conservation, but Leopold did not make this mistake. He acknowledged that financial constraints will always be a limiting factor in land management. Some conservation practices, no matter how appealing, are simply not affordable and cannot be justified. Consider the cost of restoring diverse native grassland on an old cropland field that has grown up in thick mesquite or huisache. No doubt this is a desirable practice, but when you consider the price tag, well in excess of five hundred dollars per acre, it is impractical for most operations, and the level of risk is considerable. Even with government incentives, it is difficult to justify many worthwhile conservation projects.

Chip Merrill taught his TCU Ranch Management students a two-question test to determine whether a practice was feasible on a ranch. Both questions must be answered yes in order for the proposed practice to be adopted: Is the practice ecologically sound? Is the practice economically sound? By thinking through both the ecological and economic consequences of a practice, the landowner or manager can make decisions that are good for the land and good for economic sustainability. The economic realities of conservation and land management are much more challenging than the ecological component. It is relatively easy to come up with an ecologically brilliant management plan that will improve wildlife habitat, plant diversity, livestock productivity, and watershed condition. The hard part is developing a realistic plan that is economically feasible.

The economics of cattle ranching are sobering. Today a weaned calf sells for about two to three times what it brought in 1980. In that same time period, the cost of a basic ranch pickup truck has increased at least fivefold. The costs of land, labor, feed, fuel, materials, and machinery have all far outpaced the income from cattle, making profitability extremely challenging.

Fortunately, in Texas, hunters are willing to pay handsomely for the opportunity to hunt on well-managed land. It is a great partnership that meets the demand for quality hunting and contributes substantially to the financial stability of ranches. The value of hunting has increased steadily over time and has more than kept pace with inflation, and demand remains high. For many landowners, lease hunting is an economic saving grace. Whether you personally hunt or not, we can all be thankful for hunters who are helping bear the costs of conservation and maintaining healthy lands.

It is a fine thing that Texas ranchers are paid for the hunting and outdoor recreation they provide as well as the livestock they produce. But, back in his day, Leopold lamented that "most members of the land community" had "no direct economic value," and he considered this a weakness in the conservation system of the time. This includes songbirds, predators, rodents and other small mammals, insects, and reptiles as well as the natural processes that take place on private land and benefit society.

The old ways of gathering cattle are still used on some ranches. Whether using old ways or new ways, keeping working lands intact, productive, and profitable is important, not just for the landowner but for all of society.

Responsible landowners provide a great deal of public benefit beyond the crops, livestock, timber, and hunting they produce. We see the beginnings of some landowners being compensated for the carbon that is stored in their soils. Will landowners someday be compensated for maintaining healthy watershed conditions that help keep creeks and rivers running clean or for every acre foot of water that flows into aquifers beneath their land? Will landowners receive payments for maintaining or enhancing biodiversity? There are costs and effort associated with each of these "ecosystem services"; they do not happen automatically. If society benefits from these services, should society also bear the costs?

Leopold was far ahead of his time and prophetically accurate when he said, "Conservation will ultimately boil down to rewarding the private landowner who conserves the public interest."[13] When landowners are justly compensated for the ecological benefits they provide to society, conservation will be accelerated. Conservation and economics always go together.

6

LESSONS IN
LAND MANAGEMENT

The lessons in this chapter take the next logical step beyond a study of ecology, land ethics, and basic conservation. In these lessons are found the integration of ecology, ethics, and conservation with a focus on the way that land is actually managed. Land management is necessarily a hands-on job and cannot be done simply by contemplation, knowledge, and sitting at the desk. Land managers must have a strong work ethic as well as a strong mind and a well-developed land ethic. Managing land is hard work requiring physical, mental, emotional, and financial exertion. These lessons provide a glimpse of how this is done.

A Delightful Avocation

What more delightful avocation than to take a piece of land and, by cautious experimentation, to prove how it works? What more substantial service to conservation than to practice it on one's own land?[1]

Aldo Leopold, 1932

In the article from which the above epigraph is taken, Leopold describes some ways in which a hundred-acre tract in the Midwest with one or two coveys of quail can be manipulated to produce perhaps six coveys. The article was submitted to a popular hunting magazine but was rejected by the editor as irrelevant. In those days, when a piece of land was lacking game birds, the normal remedy was to release pen-raised birds for shooting rather than managing the habitat. Leopold referred to this as "chicken-wire slums" and considered this remedy very inferior to producing wild quail in natural habitat.

In the article, Leopold describes the landowner as a "botanical and zoological engineer" who "lubricates the engine we call Nature" rather than resorting to artificial methods of propagation. By combining the skills of observation, common sense, innovation, and hard work, land engineers can replicate with their management the ideal mix and arrangement of grass, weeds, and brush for quail habitat across the entire property.

Dale Rollins says the same thing in a different way: "Visualize the honey holes for quail on your ranch—those places where you can always go and find a covey. Note what is unique or special about those areas, and then go to work with your management to 'cut and paste' those features across the entire ranch." There is infinite variation of how best to use the basic tools of conservation to manage land, and many of the best combinations may remain to be discovered.

Leopold was postulating that each piece of land, to some extent, is an experiment station, with the landowner trying various techniques to "lubricate Nature" in an attempt to increase the variety and abundance of plants and animals according to their own individual objectives. This requires the willingness and creativity to experiment with different unproven methods and practices to see what works and what does not.

Hal and Amy Zesch, who own and operate a family ranch near Fly Gap, Texas, are good examples of this kind of experimentation. The Zesches knew that prairie dogs once were keystone rodents in the region prior to their eradication many decades ago. They knew that prairie dogs have specific ecological functions and that they literally create habitat for many other species, including burrowing owls and badgers. After some initial setbacks and much effort, they have now successfully reintroduced prairie dogs to their ranch. By experimenting, failing, adjusting, persisting, and then succeeding, they now have a self-sustaining population that is expanding. They are trying to do the same thing with big bluestem, a native tall grass that has become nearly extirpated from the region. Their dream for the ranch is to help restore more of the original biodiversity, character, and productivity of the land and to pass their stewardship dreams along to their children, grandchildren, and beyond.

Most landowners are naturally curious and eager to try new things on a small scale. Without having to invest a lot of time or resources, landowners often discover a new twist to an old practice or invent an entirely novel approach. Nearly any farm or ranch has a small area where new things can be tried and evaluated. A little bit of mad scientist blood flows in the veins of most landowners as they seek better and better solutions. Many of the best techniques and methods used today began on an obscure corner of land on the backside of a field or pasture.

Collectively, professional conservationists and land management advisors have a confession to make. Most of us do not own land, and most of us do not directly carry out land management. The ideas that we pass along to others are mostly those things we have learned from landowners. When we observe something done on a ranch that works, we remember it and pass that idea on to others. When we see something that did not work, we remember that also.

Although we may have technical training in ecology and land management, most of our expertise comes after our schooling, observing what works in the real world. When we recommend something that we think will be beneficial, most of the time it is something we have seen or heard about that worked well on another piece of property. In a sense, we are plagiarizing the good ideas of others and passing them along. Landowners are a generous group—they are usually happy to pass along the discoveries from their experimentation in hopes of helping someone else. The biologist, specialist, or conservation agent is, in most cases, simply the person who relays the information.

The good news in Texas is that the avocation of good land management can also be profitable. Hence, one's vocation and avocation can be one and the same. This is especially true when wildlife management and lease hunting are closely linked to agriculture.

On many private farms, ranches, and forestlands, these are the places where all of this comes together—agriculture, habitat management, land stewardship, and the tradition and culture of hunting and sportsmanship. When done in a skillful, thoughtful way, all of these things are not only compatible but also complementary and synergistic. Landowner experimentation and the discovery of new methods of conservation and management are alive and well.

Self-Portrait

The landscape of any farm is the owner's portrait of himself.[2]

Aldo Leopold, 1939

It is impressive to consider how just a few words can convey so much truth. In the epigraph above, Leopold says that the landscape of a farm, ranch, or forest is a true reflection of the owner's personality and values.

The kind and quality of artwork displayed in someone's home or office says something about the owner. Likewise, the artwork inscribed on the land reveals much about the owner or manager of that land. All landowners create a kind of self-portrait with the type of management they apply to the land.

Two essential skills are necessary for the successful artist, and they must be used in the right combination and balance. First, there must be technical skill in the chosen art form—watercolor or oil painting, wood carving, bronze sculpture, et cetera. This may take years of practice as well as specialized training. Second, there must be vision and creativity and inspiration for the expression of ideas into art. Artists must be able to picture in their head what

Often underappreciated is the beauty and productivity of the Texas Panhandle where the High Plains meets the Rolling Plains. Here healthy, well-managed grasslands paint a picture of land health that reflects the values and skills of the landowner.

they desire to paint on the canvas. Then they must possess the technical ability to replicate that vision on the canvas or the block of wood.

Likewise, to manage and sculpt the landscape of a farm or ranch requires technical skills. There must be the ability to do all of the laborious work required, such as building fence, maintaining water, managing livestock, managing hunters and employees, planning and conducting habitat and range management work, dealing with weather extremes and market conditions, and a hundred other tasks. But there must also be creativity, vision, and individuality to do the work skillfully and in a way that is pleasing, rewarding, profitable, and lasting.

Hard work and technical conservation skills alone are not enough to create a masterpiece from a raw piece of land. The landowner or manager must also have vision, creativity, and a deep personal attachment to the land. Without this personal connection, the work required to manage the land would be overwhelming, uninspired, and discouraging. Landowners must first have a vision for the farm or ranch, what they see it looking like, and then work diligently and skillfully to make the vision a reality.

If the landscape of a farm is a reflection of the owner, we might think that such a painting is a large complex mural rather than a simple portrait. The etching, painting, and carving that is done on well-managed land is more intricate than the finest detailed piece of art. It took Michelangelo more than four years to complete the painting on the ceiling of the Sistine Chapel, and it has been preserved for over five hundred years. It is a finished masterpiece. But artwork done on the land is different. Artwork on the land is never completed. It is always in progress, it is always changing, and it always needs touchup. Artwork on the land can never be preserved. It must be continually managed.

The abundance and diversity of wildlife on any piece of property is tied directly to the owner's values and practices. If the owner desires maximum livestock grazing above all else, there is likely to be a scarcity of weeds, a lack of brush, and little wildlife, and the artwork will be more simple. If the owner wants a balance of grazing land and songbird habitat, the portrait will look much different and be more complex. A masterpiece will require great skill and creativity.

Stewardship and conservation combine both artistry and technical skill to create and maintain a landscape that is pleasing to the owner, ecologically sound, economically sustainable, and beneficial to others. The visible landscape, to some extent, does reflect the personality of landowners—their values, their likes and dislikes, and what is important to them.

What does your land reflect about you?

The Practitioner of Conservation

It has always been admitted that the several kinds of conservation should be integrated with each other, and with other economic land uses. The theory is that one and the same oak will grow sawlogs, bind soil against erosion, retard floods, drop acorns to game, furnish shelter for song birds and cast shade for picnics; that one and the same acre can and should serve forestry, watersheds, wild life, and recreation simultaneously. . . .

The plain lesson is that to be a practitioner of conservation on a piece of land takes more brains, and a wider range of sympathy, forethought, and experience, than to be a specialized forester, game manager, range manager or erosion expert in a college or a conservation bureau. Integration is easy on paper, but a lot more important and more difficult in the field than any of us foresaw.[3]

Aldo Leopold, 1934

Of all the qualities we see in Aldo Leopold, one of his finest attributes is that he was able to see the big picture. And because he left behind a wealth of writing, he still helps today's conservationists see a bigger, clearer picture of the land.

There are several worthwhile lessons in Leopold's assertion about the need for integration of conservation. The first is that conservation should not be separated from real-world economics. This is especially significant in a private lands state like Texas, where only about 4 percent of the land is public and where the private landowner must bear significant cost and financial risk in taking care of their land.

Next, Leopold reminds us that that every plant has multiple values. Grass is not just for grazing. Shrubs are not just for berries, cover, or browse. Every acre of land has multiple purposes and benefits. We may view a pasture as providing quail habitat and cattle forage, but, if it is properly managed, it is also providing watershed protection, groundwater recharge, biodiversity, songbird habitat, pollinator habitat, carbon sequestration, and other vital natural services.

The oak tree described by Leopold has many simultaneous uses and values. Some uses, such as lumber, have direct economic value. Some values are very beneficial to society but currently have no direct economic value, such as the binding of soil and slowing of floods. Other values are good for landowners, hunters, birdwatchers, and picnickers, and these can sometimes be translated into economic value. Who ever thought that an oak or any other kind of tree could be so important to so many different people?

The third lesson is that conservation is a complex endeavor, requiring more skill than any of its component disciplines alone do. A proper view of conservation requires practical knowledge, skill, and experience in soils, plants, range management, wildlife management, livestock management, economics, estate planning, and other core disciplines. The term that is sometimes used to describe this is "holistic management," which attempts to integrate the whole rather than just study or manage the parts.

The final lesson is that all of this is easy to talk about and put on paper but difficult to actually apply on a farm, ranch, or forest. There are many people who know the theory and principles of land management, but landowners or land managers are the ones who must make it work at the practical and economic level.

Leopold owned a small sandy-land tract in South Central Wisconsin. It was here that he discovered and validated many of the principles of land and wildlife management. It was here that he distilled many of the philosophical beliefs that later became his set of land ethics. Owning a piece of land and learning how to take care of it is by far the best teacher of conservation.

The greatest conservationists of today are not the scientists, professors, agency specialists, or leaders of conservation organizations. All of these perform a valuable service, but each is only a small cog in the larger machine. The real practitioners of conservation are the men and women who manage the lands, waters, and wildlife and do so in thoughtful, skillful, and responsible ways. The benefits of this include both economic gain and long-term sustainability, all driven by the kind of land ethic taught by Leopold. These are the people who produce and foster the food, fiber, wood, wildlife, water, and recreation that we all depend on. We owe them a debt of gratitude.

Your Signature

A conservationist is one who is humbly aware that with each stroke [of the axe] he is writing his signature on the face of his land.[4]

Aldo Leopold, 1949

All landowners inscribe their signatures on the land, as seen in the way they manage and treat the land. The way we sign our name to the land tells a lot about us, and it tells others what kind of conservationist we are.

When it comes to handwriting, everyone has a different and distinctive signature. Because no two signatures are the same, they help identify who we are. Likewise the signatures we write on the land will each be different and unique. There is a wide range of signature types that represent good conservation. Some prefer more open grassland for livestock grazing, while others like plenty of moderate to thick woods and brush for wildlife. Many landowners strive for a balance of these, and each has their own idea of what the right balance is.

In saying that a conservationist is one aware of signing a signature on the land, Leopold was speaking specifically of the selective thinning of wooded areas in order to create more favorable habitat. Leopold removed or reduced certain species of lesser value in order to promote other species with greater value. This practice is very much like our modern-day concept of brush sculpting. Leopold advocated a careful, measured approach to managing brush and trees to achieve specific land management goals. The concept has been adopted by many Texas landowners with beneficial results. But some have not yet heard or heeded the message.

A well-meaning landowner in West Texas purchased five thousand acres. The gentleman had been very successful in business and wanted a ranch on which to enjoy wildlife and hunting and to raise cattle. He was ready and willing to do whatever was recommended to make the ranch better. The ranch was great in many respects, except that it had extremely thick pricklypear. The new owner asked the local conservation agent what he should do about the pricklypear infestation and was told to aerial spray it. Without checking into the side effects of the herbicide, the landowner inadvertently killed most of his hackberry, sumac, wolfberry, and many kinds of forbs, all of which are important for wildlife. The damage was severe and obvious for many years. He left an unfortunate signature on the land even though his intentions were sincere and good.

Leopold included the quality of humble awareness as a vitally important characteristic for a conservationist. Awareness is the quality of constant alertness and attentiveness to what we are doing, being always mindful and cautious to ensure that our actions are achieving the intended results. Humility is the acknowledgement that there is much we still do not know. It means we are always learning and always seeking to better understand the land and improve our management. Humble conservationists admit when they make mistakes and are wise enough to learn from them.

There are many good conservation signatures being written across Texas but also some poor ones. More and more ranchers use conservative, flexible stocking rates and rotational grazing, yet overgrazing is still a problem in some places. Many landowners are conscientious about reducing excessive numbers of deer and exotic ungulates, but overpopulations are still a common concern. Many landowners now appreciate the value of weeds and forbs for habitat, but excessive weed control creates a sterile environment for wildlife on some farms. More and more landowners are working hard to restore native grasslands, but exotic grass monocultures are still being planted on some ranches.

The worst land signature of all is the splitting up of large tracts of land into many smaller tracts. Fortunately, a growing number of Texas landowners are establishing conservation easements to maintain large intact landscapes, and some are going one step further by amalgamating small adjacent tracts into larger properties.

Our modern tools are now more sophisticated and powerful than the axe. They can do greater good or greater harm, depending on how they are used. We need humble awareness more than ever in the way we use these tools to write our names on the land.

What kind of signature are you writing on your land, and what does it say about your brand of management? Does it show that you are careful about the treatment of soil, water, plants, and animals, or does it demonstrate carelessness or apathy? Does your signature provide positive or negative advertising for conservation and private lands stewardship? People are watching. One of our jobs is to be good conservation ambassadors, showing the public that we are taking our stewardship responsibilities seriously.

Overgrazing

Overgrazing is more than mere lack of visible forage. It is rather a lack of vigorous roots of desirable forage plants. An area is overgrazed to the extent its palatable plants are thinned out or weakened in growing power. It takes more than a few good rains, or a temporary removal of livestock, to cure this thinning or weakening of palatable plants. In some cases it may take years of skillful range management to affect a cure; in others, erosion has so drained and leached the soil that restoration is a matter of decades.[5]

Aldo Leopold, 1933

Leopold's description of overgrazing demonstrates his keen understanding of grass growth, livestock grazing, and range management. Overgrazing is often considered one of the deadly sins of both rangeland and wildlife management. Since the effects of overgrazing can be so harmful, it is helpful to understand the different ways that the term "overgrazing" is used. Although it can be a serious problem, overgrazing is a misunderstood concept and can be misdiagnosed.

There are two types of overgrazing. The first and less serious type is seen in a pasture that is grazed short or, as Leopold says, has a "lack of visible forage." This type of overgrazing is a short-term condition that is easily reversible with management and rainfall. This type of overgrazing can result from careless management, but it can also happen under good management. For this reason, it is hard to judge overgrazing by a single observation or by driving by and looking over the fence.

For example, a rancher may use rotational grazing where a herd of livestock is placed in a pasture for a period of time. By the end of the grazing period, especially when conditions are dry, the pasture may be grazed short and may appear to the untrained eye to be overgrazed. However, after the livestock are moved to other pastures, the recently grazed pasture will have perhaps six months or more of recovery time, in which the grasses and other plants will have ample opportunity to regrow if it rains. There is no lasting damage to plants under this kind of grazing, especially where prudent managers are constantly evaluating grass growth and rainfall and making needed adjustments in livestock numbers.

However, if this type of short-term overgrazing persists without adequate recovery periods, it progressively becomes more serious, with long-lasting consequences. This is the type of overgrazing described by Leopold. This long-term overgrazing seriously harms the health of plants, especially the root system. When plants are continuously grazed and re-grazed, the plant

Chronic overgrazing exacerbated by drought is debilitating. The effects are crippling and long-lasting. Decades of careful management are needed to restore land damaged in this way.

is forced to draw on stored energy reserves to produce new growth. Excessive and frequent removal of green photosynthetic leaf tissue drains the plant's essential carbohydrate reserves, damaging the root system.

When the root system is diminished by overgrazing, plant growth is reduced and plants may go into a state of semi-dormancy to survive. Such plants, injured and stressed by overgrazing, may survive in a weakened, stunted, and unproductive state for many years. Unless grazing management is changed, these plants eventually die due to chronic long-term overgrazing, and it often happens during drought.

In a pasture composed of many plant species, overgrazing happens at different stages for each species in a somewhat predictable manner. The more palatable plants are the first to be eaten and may be grazed short and re-grazed before other grasses are grazed at all. This selective and preferential grazing is the reason why the better grasses are gradually thinned and eventually eliminated from many pastures. Simultaneously, the lesser grasses and unpalatable weeds may be left un-grazed, allowing them to increase.

To the trained eye, this kind of chronic overgrazing is easy to discern. For example, a Central Texas pasture dominated by red grama, Texas grama, hairy tridens, curly-mesquite, and threeawn is a clear indication of long-term overgrazing. The absence or scarcity of little bluestem, Texas cupgrass, plains lovegrass, vine mesquite, and other desirable grasses will verify the diagnosis. The cure for this kind of long-term overgrazing is not easy or quick. Leopold stated that it takes years or decades of skillful range management to reverse the damage caused by long-term overgrazing, but he may have been too optimistic regarding the time frame of recovery, especially for semi-arid ranges.

Most of Texas was subjected to long-term sustained overgrazing from the late 1800s through the mid-twentieth century. This widespread, persistent overgrazing was due to a combination of apathy, ignorance, and unrealistic expectations. It was a sad era in our history of natural resource management. The ecological and economic damage incurred was immense and long-lasting. Fortunately, with an increase in our knowledge of range management and a parallel improvement in stewardship ethics, Texas rangelands are in much better condition than they were in Leopold's day. Most of today's ranchers are more careful, conscientious, and conservative than their forebears. Hopefully, in another generation or two, examples of abusive overgrazing will become uncommon and difficult to find.

The Solution to Overgrazing

It has been asserted that erosion is the result of overgrazing, and that some local overgrazing is difficult to avoid, even on ranges that are not overstocked. But nobody advocates that we cease grazing. The situation does not call for a taboo upon grazing, but rather constitutes a challenge to the craftsmanship of our stockmen and the technical skill of grazing experts in devising controls that work. . . .

The stockmen must recognize that the privilege of grazing use carries with it the obligation to minimize and control its effects by more skilful and conservative methods.[6]

Aldo Leopold, 1924

Leopold's comments about overgrazing were prompted by the active erosion taking place on Forest Service lands in the Southwest in the 1920s. Leopold was very much concerned about soil erosion. Of all of the natural resource issues he was involved in, the loss of topsoil was perhaps his greatest concern. Unlike many other natural resource problems, soil erosion represents an irreversible loss of productivity. Erosion affects our food production, water supplies, wildlife populations, and the overall productivity of the land. It is no wonder Leopold wrote so extensively and with such great conviction on this subject.

On rangelands, a healthy cover of grass is what holds the soil in place and controls erosion. Land blanketed by grass, with healthy roots and a layer of litter on the surface, is protected from erosion. Overgrazing thins the protective blanket and weakens the roots, leaving the soil exposed and vulnerable to erosion. Unfortunately, many Texas ranches have experienced extreme soil erosion due to overgrazing over the last 140 years.

The removal of livestock has often been advocated to correct the damage caused by overgrazing. It is frequently recommended by wildlife biologists, ecologists, and botanists. Perhaps they see the damage caused by longtime overgrazing and reach the seemingly logical conclusion that removing livestock will correct the problem. Their intentions are good, but their recommendation is short-sighted. Although improper livestock grazing has undoubtedly caused great harm to ranges, watersheds, and habitats, Leopold did not advocate the long-term exclusion of grazing.

Why would Leopold be reluctant to call for removing livestock from eroding, overgrazed rangeland? The answer is that Leopold always tried to see the big picture. Although he was an ecologist and wildlife manager who detested land abuse, Leopold was also a realist. He understood the economic and cultural components of ranching just as much as he did the ecological aspects.

If livestock are permanently removed from the land, there are immediate and long-lasting impacts on the economics of the land and the people who own or manage the land. Leopold seemed to understand the importance of productive and profitable agriculture and the importance of keeping people living on the land. Without livestock, and without the people who tend and manage the livestock, it is more likely that ranches will slowly revert back to neglected, unmanaged land.

There are some valid reasons for the occasional short-term removal of live-stock, especially in response to extreme drought or wildfire or to jump-start the recovery of overgrazed range. Many ranchers and range managers have seen the wisdom of temporarily suspending grazing for several years to hasten the recovery of degraded riparian areas or other sensitive lands. However, in the majority of cases, as Leopold noted, long-term exclusion of livestock is not the answer to overgrazing or soil erosion. Instead, the better answer usually involves management—the constant monitoring of conditions, frequent adjustment in livestock numbers, and a planned method of rotational grazing where each pasture is getting needed rest and recovery every year.

Intuition and common sense may indicate that if serious livestock overgrazing is occurring, the simplest and most obvious solution is to stop grazing. However, simple solutions are not always the best solutions. As Leopold stated, the solution to overgrazing is twofold: greater skill on the part of grazing management specialists and a greater sense of obligation on the part of the range manager.

Fortunately, since Leopold's day, the science and art of range management has made many important advances, and these are slowly being incorporated into many ranching operations. Furthermore, many of today's ranchers have a deeper sense of ethical obligation to be responsible caretakers of the land. Overgrazing is still a problem in Texas, but fortunately it is less common than it used to be. We are headed in the right direction.

The stewardship of Texas lands is a high calling. Doing this in a manner that is economically and culturally sustainable is a great challenge.

Grazing and Erosion

I have stated that any system of grazing, no matter how conservative, induces erosion. The proof of this statement . . . may be seen almost anywhere in the hills.[7]

Aldo Leopold 1921

The above quotation on grazing and erosion should seem familiar. It appeared in chapter 4 in the lesson titled "Dogmatic Statements." There the danger of making bold, rigid statements was discussed, but in this lesson the erroneous statement that grazing causes erosion is addressed. Like many false or misleading statements, there is usually some truth mixed in with error. Grazing done the wrong way can cause erosion, but grazing done the right way can heal erosion.

It is still true today that the manner of grazing on some ranches induces soil erosion. Cow trails can concentrate runoff and form gullies. Heavy use near water, salt, mineral, and feeding areas often results in sparse grass cover and erosion. When livestock stay too long in a pasture, the grass cover may be insufficient to prevent erosion. These common examples of grazing-induced erosion are not difficult to correct if the manager has the desire. But there are other aspects of grazing-induced erosion that are more difficult to overcome.

The worst enemy of healthy rangeland is using set stocking rates and refusing to reduce animal numbers in drought. Every rancher knows that rainfall and forage production varies greatly from year to year, yet some producers keep the same number of animals regardless of forage production. Overgrazing and erosion is inevitable with this kind of management.

Another common problem that often leads to heavy grazing and erosion is adherence to traditional but outdated stocking rate guidelines. For example, many livestock producers in Central Texas have been told that the area will support one cow per twenty or twenty-five acres on native rangeland. In reality most ranches will not support those numbers, and the result is short-grazed pastures and subtle, cumulative erosion.

The good news is that overgrazing and erosion have been much reduced over the last fifty years. Our knowledge of grazing management has come a long way since Leopold made his bold and incorrect statement. Furthermore, the ethics of good land stewardship that drive management decisions are now more well developed.

We now know a lot more than earlier ranchers knew about sustainable grazing. We know that chronic heavy grazing stunts the root system and hurts the vigor of grass plants, which leads to a decline in grass cover. We now know that grasses can be substantially grazed, but then they need an adequate period of time to recover before being grazed again. We know the importance of maintaining residual grass and grass litter to protect the soil and improve rainfall penetration.

By using flexible stocking rates and adaptive rotational grazing, grass cover will improve and erosion will be curtailed. Areas of active erosion can often be healed by the right kind of grazing management. Denuded areas can usually become covered by desirable grasses, although the process may be slow. Soil health and productivity will improve as grass and litter cover increases. Deeper-rooted grasses will gradually increase, taking the place of shorter grasses. What was damaged by the previous generations can be slowly restored.

New forms of grazing with new names are being actively promoted, and if done correctly they can produce good results. But the new intensive methods are not for everyone, and even simple forms of grazing management can be successful in improving a damaged grass cover. The positive results that can be achieved through grazing management are usually commensurate with the skill of the operator and the effort invested.

Yes, overgrazing still occurs, and, when it does, it causes erosion. We have a long way to go to make overgrazing an uncommon scene in Texas. But the trend is positive, toward improved grazing management, better grass cover, and reduced erosion.

Aldo Leopold, the father of wildlife management, also had a keen interest in soil conservation and grazing management. He knew that when the soil is being taken care of, water, wildlife, timber, and grazing can be productive and sustainable.

Cleavage

Conservationists are notorious for their dissensions. . . . In each field, one group (A) regards the land as soil, and its function as commodity production; another group (B) regards the land as a biota, and its function as something broader. . . .

In my own field, forestry, group A is quite content to grow trees like cabbages, with cellulose as the basic forest commodity . . . its ideology is agronomic. Group B, on the other hand, sees forestry as fundamentally different from agronomy because it employs natural species, and manages a natural environment rather than creating an artificial one. . . . To my mind, Group B feels the stirrings of an ecological conscience.

In the wildlife field, a parallel cleavage exists. For Group A, the basic commodities are sport and meat; the yardsticks of production are ciphers of take in pheasants and trout. Artificial propagation is acceptable as a permanent as well as a temporary recourse. . . . Group B on the other hand, worries about a whole series of biotic side-issues [such as]. . . predators . . . exotics . . . shrinking species . . . threatened rarities . . . wildflowers[.] Here again it is clear to me that we have the same A-B cleavage as in forestry.[8]

Aldo Leopold, 1949

Aldo Leopold described cleavage as the separation between two ways of thinking about the land, calling them Group A and Group B. Leopold recorded these observations in part 3 of *A Sand County Almanac* and said that his was a dissenting opinion of the norms of that day: "Only the very sympathetic reader will wish to wrestle with the philosophical questions of Part III."

According to Leopold, Group A conservationists see the land in a very pragmatic way, as a means to produce something of benefit to mankind, whether it be crops, livestock, fish, wildlife, or wood products. Group B conservationists also desire for the land to be productive and profitable, but they look deeper than quantitative measures of production; they also look at qualitative measures of land health.

Leopold did not disparage Group A thinkers; after all, they are working hard to produce necessary or desirable products from the land. However,

Leopold considered Group B to be farther along in developing and being guided by responsible conservation ethics.

In our day, Group A might be exemplified by those deer managers who put the extreme emphasis on the size of antlers at any cost, using any and all methods possible. Group B deer managers are also interested in and thrilled by impressive antlers; but they are equally concerned about habitat quality, the experience of the hunt, and other wildlife species as well as the aesthetic, emotional, social, and ecological elements of wildlife management.

Keeping careful track of antler score has become the overriding concern in many hunting camps these days. More than a few landowners, hunters, and deer managers now lament this trend as they observe the effect it has on the enjoyment of hunting. When the numerical score of a buck defines the hunt, something important is missing. When a hunter, especially a young hunter is disappointed that their buck "only scores 120," something is amiss.

The A-B cleavage is also apparent in fisheries management, farming, and livestock production. Group A livestock ranchers consider maximum forage and livestock production as the primary goal, even if it means monocultures of exotic grasses and high inputs of feed, hay, fuel, fertilizer, and pesticide. Group B ranchers are also interested in production but also look at the bigger picture of native plant diversity, rangeland health, watershed condition, wildlife habitat, and long-term sustainability.

Today in Texas many fine landowners find themselves more closely aligned with Group A. This has been the primary culture of agriculture and wildlife management. However, there is also a large and growing number of Group B landowners with deep ethical connections to the land and a strong sense of ecologically based stewardship.

Group A says bigger and more is always better. Group B considers the compound rippling effects (economic, ecological, and social) of all actions before decisions are made. Group A is not necessarily wrong, and Group B is not necessarily right. It is often a subtle choice between what is good and what is best and what is the most sustainable. No matter which end of the spectrum you gravitate toward, it is always wise to count the cost and consider the side effects of everything we do or don't do in natural resource management.

These dissensions and cleavages that Leopold wrote about many years ago are alive and well in Texas conservation. In any discipline or endeavor, there are some simple, basic truths that are easy to understand and implement. There are other deeper truths and complex principles that are more difficult to grasp and require contemplation and inner struggle. The dilemmas of the A-B cleavage are something that every landowner, hunter, fisher, biologist, and nature lover must wrestle with.

With Axe in Hand

The wielder of an axe has as many biases as there are species of trees on his farm. In the course of the years he imputes to each species . . . a series of attributes that constitute a character. I am amazed to learn what diverse characters different men impute to one and the same tree.

Thus to me the aspen is in good repute because he glorifies October and he feeds my grouse in winter, but to some of my neighbors he is a mere weed, perhaps because he sprouted so vigorously in the stump lots their grandfathers were attempting to clear. . . .

The tamarack is to me a favorite second only to white pine, perhaps because he is nearly extinct in my township (underdog bias), or because he sprinkles gold on October grouse (gunpowder bias), or because he sours the soil and enables it to grow the loveliest of our orchids, the showy lady's slipper. On the other hand, foresters have excommunicated the tamarack because he grows too slowly to pay compound interest. . . .

To me an ancient cottonwood is the greatest of trees because in his youth, he shaded the buffalo and wore a halo of pigeons [passenger pigeons], and I like a young cottonwood because he may someday become ancient. But the farmer's wife (and hence the farmer) despises all cottonwoods because in June the female tree clogs the screens with cotton. . . .

I find my biases more numerous than those of my neighbors because I have individual likings for many species that they lump under one aspersive category: brush. Thus I like the wahoo, partly because deer, rabbits, and mice are so avid to eat his square twigs and green bark and partly because his cerise berries glow so warmly against November snow. I like the red dogwood because he feeds October robins, and the prickly ash because my woodcock take their daily sunbath under the shelter of his thorns. I like the hazel because his October purple feeds my eye, and because his November catkins feed my deer and grouse. . . .

The farmer who would rather hunt grouse than milk cows will not dislike hawthorn, no matter if it does invade his pasture . . . and I know of quail hunters who bear no grudge against ragweed, despite their annual bout with hay-fever. Our biases are indeed a sensitive index to our affections, our tastes, our loyalties, our generosities, and our manner of wasting weekends.[9]

Aldo Leopold, 1949

Today very few of us wield an axe when conducting brush management or tree thinning. Our tools and methods have changed, and we have adopted those that are faster and more efficient. However, the values, characters, and biases we assign to various trees and shrubs are still very much like what Leopold described. We judge their virtues and beauty by our own criteria and work to either encourage or discourage them according to our likes and dislikes. This is called management, and this is how we manipulate and shape the ranges and woods to our wishes. In making our decisions whether to keep or kill certain trees and shrubs, we consider both the reasons to keep and the reasons to remove them. In many cases we just thin certain species down to a more desired density. In so doing, we naturally impart our own personal values and preferences to our work.

For example, greenbrier is normally despised and cussed for its impene-trable growth and wicked thorns, yet every young rabbit hunter knows by experience there must be a cottontail or two in each thicket and sometimes a covey of bobwhites. The lowly, spine-laden algerita is often disparaged for taking up too much space, but songbirds are especially fond of the tart red spring berries, and the new tender twigs are relished by deer as well as by sheep and goats. The creosote bush often protects the bush muhly and fills the desert air with the most refreshing and invigorating smell, but it is mostly scorned as worthless.

The cholla with its wickedly barbed spines is cursed by the cowboy, but its yellow fruit are savored by pronghorn and mule deer, especially in drought. Mesquite, cedar, and pricklypear, all very much detested (they frequently get way too thick), are each very desirable in the right amount, providing benefits to livestock and wildlife alike.

The same prickly ash that protected Leopold's woodcock provides coverts for Rolling Plains bobwhites and plenty of seed for September doves, yet many consider it only a thorny nuisance. The catclaw mimosa is very much unloved, but its matrix of spine-loaded branches protects the remnants of little blue-stem that in turn help conceal the turkey nest. And who would argue with the splendor of scarlet sweetgum foliage in the fall? Yet it is reviled because it encroaches too quickly into open pastures in East Texas. These and a hundred other trees and bushes inhabit every region of Texas, each with their own mannerisms, uses, and character—some agreeable and some disagreeable.

Today's land stewards may no longer wield an axe to thin certain species and encourage others, but the same thoughtfulness and careful attention should be applied in our brush control activities. We practice a proper form of discrimination as we make these decisions. Should the plant be spared for its positive contributions, or should it be removed or reduced to favor something else? Leopold wrote, "A conservationist is one who is humbly aware that with each stroke of the axe, he is writing his signature on the face of the land." No matter the tools we use, may our signatures on the land be such that future generations enjoy, appreciate, and benefit from our work.

Private Lands, Public Interest

Every landowner is the custodian of two interests, not always identical—the public and his own. What we need is a positive inducement or reward for the landowner who respects both interests in his actual land practice.[10]

Aldo Leopold, 1934

Conservation will ultimately boil down to rewarding the private landowner who conserves the public interest.[11]

Aldo Leopold, 1934

Leopold was far ahead of his time. Today more and more policy makers are seriously considering how to fairly compensate private landowners for being responsible custodians of the public interest.

The general public does not know very much about wildlife management, watershed management, range management, soil conservation, forestry, or wetland management, but everyone depends on the benefits accrued from these pursuits. Every citizen, no matter how far removed from the land, relies on clean water that originates primarily on ranch lands and forests. Everyone benefits when the soil is healthy and kept in place by a good cover of grass or trees. Nearly everyone appreciates the myriad species of birds, mammals, reptiles, butterflies, grasses, wildflowers, shrubs, and trees. In Texas these all depend on private land.

The fancy name that describes the combined societal values of well-managed land is "ecosystem services." Although it is a somewhat sterile and academic term, it includes the sum of all the positive things that are produced, provided, or sustained on healthy land. Other than the tangible products and benefits of food, fiber, wood, and recreation, ranchers, farmers, and forest owners are generally not compensated for these services.

How much value can be assigned to abundant clean surface water, aquifer recharge, erosion control, biodiversity, pollinator habitat, or a dozen other essential services? Scientists are now attempting to determine the value of these less-tangible benefits, and someday there may be a system of payments or incentives to landowners who provide these services by their management.

Here are a couple of prime examples. Well-managed land helps to buffer against the extremes of drought and flood. When the water cycle is functioning properly, land absorbs and stores water for a slow and gradual release into springs, creeks, and rivers, thus providing a more sustained flow of water, even in dry times. When too much rain comes too quickly, well-managed land soaks up, slows down, and stores rainfall temporarily, thus reducing the amount of fast runoff and reducing the impacts of flooding.

Charley Christensen operates a large ranch for the Cargile family on Rocky Creek west of San Angelo, Texas. He tells the remarkable story of how their management along the creek effectively tamed a flood. The ranch had set aside what they call the riparian buffer zone along several miles of the creek to be rested from grazing for several years until the desired vegetation had a chance to develop. Within a few years, the native riparian vegetation increased, providing improved fish and wildlife habitat and protecting the banks from erosion. During one particularly severe storm, five inches of rainfall fell quickly on the upstream watershed, and the runoff began to pour off the adjacent hills and into the creek. The sound of the violent whitewater from upstream could be heard as it headed toward the ranch. The turbulent

John Graves guides an angler on the Llano River. Healthy watersheds on private land process and protect the waters that flow in publicly owned rivers, which in turn sustain healthy fisheries and provide recreation.

floodwaters gushed into the ranch, carrying great loads of soil, gravel, and rock that had eroded from the banks and streambed. Charley drove to the downstream water gap where the creek exited the ranch, expecting the worst.

What he noted was profound: "the creek was wide and smooth, and the water gap was intact." The dense vegetation along the creek had taken much of the energy out of the raging floodwaters. The riparian zone had slowed the waters and filtered and trapped much of the incoming sediment, and the waters spread out wide. The net effect was that the slower, wider waters had more time and area to soak into the alluvial water table, thus providing recharge for sustained base flows, and sediment was trapped in the floodplain rather than delivered downstream, thus improving water quality. Some tangible benefits accrued to the ranch and some less tangible benefits accrued to everyone downstream, including the population of San Angelo. Every citizen benefits from these kinds of services, even though most don't realize it is happening.

With regard to greenhouse gases and climate change, it should be reassuring to know that millions of tons of atmospheric carbon are put into the ground and stored in the soil as beneficial organic matter. Healthy forestland and rangeland are two giant reservoirs for carbon sequestration, and most of this happens on private lands—for the benefit of many.

It's a win-win combination for landowners and the public when each party benefits from the other in a mutually advantageous arrangement. The public is accustomed to paying for the commodities grown by farmers, ranchers, and forest owners. The public is not yet accustomed to compensating landowners for the other benefits they provide. But fair compensation must be fair to both sides—the landowner and the public who funds the compensation.

For many years and continuing today, a rich system of monetary payments to landowners through federal farm bill and other government-funded programs has operated. On rangeland, these incentives are mostly in the form of payments for the implementation of practices such as cross fences, water development, brush control, and grazing management. Much good work has been accomplished through these popular programs. However, too often it is merely a payment for a one-time practice without the expectation of long-term benefit. What is currently missing from these programs is the system of ongoing management that will ensure that public benefits will actually accrue and be sustained. We have some of the right ideas about compensating landowners for legitimate societal benefits, but we are still a long ways from the win-win combination described by Leopold.

One of the core messages of today's conservation community is that well-managed private land produces a sustainable, cost-effective, and abundant supply of products and services for the public benefit. Several outstanding conservation organizations are heavily engaged in educational efforts to help schoolchildren understand where their food, water, clean air and wildlife come from and why private lands stewardship is so vital. They are tomorrow's voters and policy makers. We must continue and accelerate these important efforts and be on the forefront of finding ways to fairly compensate the private landowner for conserving the public interest.

There is a fascinating side lesson here that illustrates the interwoven nature not just of the land but also among the people that manage the land. Charley Christensen, mentioned above, was a student of Chip Merrill, having gone through the TCU Ranch Management Program. The ranching and stewardship principles taught by Merrill are having an inevitable rippling effect among his former students, the lands they manage, and the people they work for and work with. Even more interesting is the fact that Charley is the grandson of Chris Christensen, who was dean of agriculture at the University of Wisconsin and the person who hired Aldo Leopold to be the country's first professor of wildlife management in 1933. Dean Christensen had the foresight and wisdom to see that successful agriculture is more than raising crops and livestock and that the culture of farming should include dividends of wildlife, water, natural beauty, and richness of living.

We can never fully chart the complex family tree of the people of the land, but it is clear that its branches spread far and wide and its roots deep.

A Land to Please Everyone

When the cows . . . were first turned out upon the hills . . . everything was all right because there were more hills than cows, and because the soil still retained the humus which the wilderness vegetation through the centuries had built up. The trout streams ran clear, deep, narrow, and full. They seldom overflowed. . . . The deep loam of even the steepest fields and pastures showed never a gully, being able to soak up any rain as it came, and turn it upward into crops, or downward into perennial springs. It was a land to please everyone, be he an empire-builder or a poet.[12]

Aldo Leopold, 1935

Leopold describes the conditions of the Coon Valley watershed in West Central Wisconsin in the middle to late nineteenth century just as it was beginning to be developed for dairy farming. It is a beautiful description of natural resource abundance, balance, and stability.

At that time, the Coon Valley watershed absorbed and stored its rainfall and snowfall; the healthy soil acted like a huge sponge. The soil had rich, inherent fertility and water-holding capacity due to the organic matter and the workings of microorganisms that kept it productive and porous year after year. The roots of trees and grasses held the soil in place, and the blanket of plant litter on the surface insulated and protected the soil. Most precipitation soaked into the ground to be used by plants or to seep below to sustain springs and feed aquifers. Creeks and rivers retained their natural character and flowed strong and clean.

But by the early 1900s, the rapidly expanding dairy herds were taking a toll on Coon Valley. The prairies were grazed off, and the once-rich timberlands were cut to make room for more pasture. By the time Leopold wrote in 1935, the land had deteriorated into something sad and unrecognizable. "Gone now is the humus of the old prairie which enabled the ridges to take on the rains as they came," he wrote. "Every rain now pours off the ridges as from a roof. The ravines of the grazed slopes are the gutters. In their pastured condition they cannot resist the abrasion of the silt-laden torrents. Great gashing gullies are now torn out of the hillside. Each gully dumps its load of hillside rocks upon the fields of the creek bottom, and its muddy waters into the already swollen streams."

No matter where we live it is good to look back in history to discover what the land looked like and how it worked prior to settlement. Across most parts of America, the late 1800s and early 1900s were an era of overuse and natural resource degradation. The Norwegians who settled Coon Valley had to make a living, but they did not understand ecology, conservation, or land ethics. Similar stories can be repeated in every region of every state. The lessons learned should keep us from repeating the mistakes of the past.

The Coon Valley watershed later became one of the first cooperative watershed conservation projects in the country. The Soil Erosion Service, which later became the Soil Conservation Service, embarked on an ambitious program of technical and financial assistance with willing farmers. It was a voluntary partnership between conscientious landowners and locally led government assistance. It did not happen overnight, and there were many mistakes and setbacks, but little by little the grasses returned to hold the soil in place and capture the rainfall. The steeper slopes were returned to trees. The creeks were fenced to allow protection and proper grazing of streamside vegetation. Today Coon Valley is home to blue-ribbon trout streams, stable hillsides, productive agriculture, and natural beauty.

Leopold knew that land had to be managed and worked to produce goods, services, and income. He recognized that land management is not just about ecological integrity but also about economic and social well-being. This is sometimes referred to as the "triple bottom line," whereby financial, ecological, and societal benefits all accrue simultaneously and for the long haul.

Well-managed land is the proverbial win-win for everyone, whether rural residents or city dwellers. It is beautiful enough to inspire poets and productive enough to generate returns for the owner, which in turn sustain families and communities. Along with these benefits are good wildlife, good fish, and good water, all perpetually renewable. These are the results of genuine stewardship.

7

LESSONS IN
WILDLIFE MANAGEMENT

The lessons in this chapter focus on one particular aspect of land management—management aimed at the production of wildlife. Most landowners have an interest in maintaining or increasing the kinds and numbers of wildlife on their land. Some are primarily involved in production agriculture, with a secondary interest in wildlife, while others have a primary objective of producing and managing wildlife. In some cases significant income is generated from hunting or other wildlife-related enterprises, but in nearly all cases, whether for profit or pleasure, landowners generally derive much personal enjoyment and satisfaction from having good wildlife populations. These lessons provide direction and inspiration to guide and motivate effective wildlife management. One will quickly see that the previous lessons in ecology, ethics, and conservation are intertwined with successful wildlife management.

More Than a Vague Liking

To accomplish [the] complex adjustments (which we may call collectively "wildlife management"), the farmer must be moved by something more than a vague liking for wild things. He must be moved by a positive affection for the fauna and flora as a whole and he must take pride in the skill and knowledge exercised in their management. In short, each farmer must build up, and cherish, his social "rating" as a producer of wild as well as tame animals and plants.[1]

Aldo Leopold, 1941

Leopold summarizes five important aspects of successful wildlife management—complexity, affection, knowledge, skill, and pride. First, he acknowledges that wildlife management is a complex discipline. There is nothing simple or easy about it—the landowner is constantly adjusting management to take the land in the desired direction for whatever species one is interested in. Sometimes it involves coarse, heavy adjustments and always constant fine-tuning and tweaking. There is no simple recipe explaining just how to do it since every piece of land is different, every year is different, and each landowner has unique goals and capabilities.

The complexities of management are nearly infinite: adjusting livestock numbers, as well as the timing and movement of grazing; adjusting deer numbers, sex ratios, and age structure through hunting; adjusting the ratio of grass, forbs, and brush with grazing, fire, mechanical methods, or herbicides. All of this is done with the sober realization that weather often trumps even our best efforts of management. Wildlife management is truly complex.

Second, Leopold insists that the landowner must be moved by a genuine affection for wild plants and animals. The landowner must literally love the land in order to be a successful manager. Landowners may or may not verbally express their inner emotions and affections, but it is clear that most of the good managers have a strong and loving connection to the land. Leopold points out that this affection is for the land as a whole, not just the deer or birds or flowers or grass but for the whole intricate tangle of nature. One can go through the motions and mechanics of land and wildlife management, but without affection and appreciation the results will often be disappointing and superficial.

Those engaged in wildlife management must be moved by something more than a vague liking for wild things. An up-close encounter with a mature, wild South Texas buck is something you will always remember, whether you hunt or not.

Motivated by this affection for the land, Leopold adds that knowledge and skill are also needed. It is not enough just to love the land with the heart; there must also be the mental and physical knowledge and skills. Knowledge can be gained at seminars, school, and from magazines, books, and other places. This includes knowledge about soil, water, plants, animals, ecology, machinery, methodology, and a dozen other topics. But that knowledge must translate from the head to the hands, putting raw information into practice with hard work, determination, creativity, and skillful management. Skill involves putting knowledge to practical, beneficial use.

Finally, Leopold describes the value of taking pride in what the land produces. Most good farmers are proud of their crop yield, whether it is bales of cotton, bushels of wheat, or tons of hay per acre. Most good ranchers are proud of their productivity ratings, such as calf crop, weaning weight, gain per head, or pounds of beef per acre. Wildlife managers likewise are proud of their measures of success, which might include coveys flushed, covey size, poults per hen, songbird diversity, fawn crop, antler development, or other such metrics.

But in addition to these more traditional measures, the successful wildlife manager also takes pride in other things. Perhaps you have the only colony of big bluestem within twenty miles or one of the few populations of mountain mahogany, and these are thriving and increasing under your management. Maybe it is the pride in knowing that you have nesting eagles or falcons that no one else is aware of. Or perhaps your place is home to some rare flower, lizard, or fern that you are taking special measures to protect. Even if you do not have something rare or unique, you may be proud that you have been able to restore the old worn-out cropland field to an impressive diversity of native grasses and forbs. Pride in such things is a good quality when it congratulates or motivates good management. These things are some of the raw ingredients of true stewardship and good wildlife management.

Leopold's Repentance

I personally believed . . . in 1914 . . . that there could not be too much horned game, and that the extirpation of predators was a reasonable price to pay for better big game hunting.[2]

Aldo Leopold, 1945

As an avid hunter and confused ecologist, young Aldo Leopold was an outspoken proponent of intensive predator removal early in his career. Leopold knew the obvious—that a decrease in predators would mean an increase in big game. He advocated the aggressive and widespread eradication of predators, especially wolves and mountain lions in the Southwest.

Young people, full of confidence and short on experience, often overestimate their understanding of a subject and make hasty proclamations and unwise decisions. Who among us has not done this? As we mature and gain experience, the dogmatic views we once held dear sometimes evolve and our opinions mellow. Notice the evolution of Leopold's thoughts about predators from 1919 to 1939:

1919—"The advisability of controlling vermin is plain common sense, which nobody will seriously question."[3]

1920—"To try and raise game in a refuge infested with mountain lions, wolves, coyotes and bobcats, would, of course, be even more futile than to try and run a profitable stock ranch under similar conditions. Predatory animals are the common enemy of both the stockman and the conservationist . . ."[4]

1920—"It is going to take patience and money to catch the last wolf or lion in New Mexico. But the last one must be caught before the job can be called finally successful."[5]

1930—"Future predator control must be localized and discriminate."[6]

1939—"Perhaps there are too many deer . . . Perhaps it was a mistake to clean out the wolves."[7]

1939—"Browsing animals . . . are in constant danger of destroying their own range . . . hunting alone is seldom a sufficiently delicate control to keep the herds in balance. We need predators as well."[8]

"Repentance" is mostly a religious term. It means to change one's mind about something, with a corresponding change in behavior. But repentance does not always happen in a moment. Over the course of many years, Leopold changed his views about predators and big game populations, and he repented of his earlier position.

Toward the end of his life, Leopold reflected back on his views about predators. In his famous essay "Thinking Like a Mountain," published in 1949, Leopold recalled a 1909 encounter with wolves along the Arizona–New Mexico border: "In those days we had never heard of passing up a chance to kill a wolf. . . . I was young then, and full of trigger-itch; I thought that because fewer wolves meant more deer, that no wolves would mean hunters' paradise. . . . Since then I have lived to see state after state extirpate its wolves. I have watched the face of many a newly wolfless mountain, and seen the south-facing slopes wrinkle with a maze of new deer trails."[9] He went on to note that the excessive deer population, unchecked by predators, was literally destroying their own food supply by over-browsing.

Leopold changed his mind about widespread and intensive predator removal as wise game management policy. However, Leopold's repentance does not mean he condemned all predator management. While he changed his thinking about predator eradication, there is no indication that he was opposed to the thoughtful management of predators in specific cases where it was warranted.

Although the ecological role of predators is now better appreciated, predator control still has a legitimate place in ranching and wildlife management. Ask any experienced rancher in pronghorn country, and they will tell you that excessive coyotes and bobcats are a major factor in low fawn crops, thus keeping antelope numbers suppressed below carrying capacity. Predators are not the only factor, but often they are the most important. Likewise, where white-tail or mule deer fawn survival is low year after year, coyote reductions prior to fawning season will often yield good results. Clearly, good habitat management is the cornerstone of wildlife conservation, but predator control can also be sound practice in some cases.

Brown-headed cowbirds wreak havoc on songbird populations. While not technically a predator, the effect is exactly the same. Trapping and killing native cowbirds by the thousands has become one of the primary tools used to increase songbird populations and is widely accepted as good wildlife management. The killing of one species to benefit another is the essence of predator management. It can have a legitimate place for livestock production, game production, and songbird production.

Predator control is still a controversial topic chockful of emotion. It is good to remind ourselves that predator management is simply a tool. It is neither inherently good nor inherently bad because, like any tool, it can be used properly or improperly. The skill with which any tool is used in each individual situation will determine whether it has a positive or negative impact. Let us not repeat the early mistakes of young Leopold, but neither let us discard a valuable management tool just because it was previously misused.

Out of Kilter

Every woodsman knows that deer in many places are exterminating the plants on which they depend for food. Something is out of kilter.[10]

Aldo Leopold, 1939

A woodsman is one who spends enough time in the woods, hills, pastures, and river bottoms to see and understand what is happening. A woodsman notices subtle things overlooked by others. Not content to view the world through the windshield, woodsmen walk the pastures, or ride horseback, pausing frequently to observe, ponder, and study. In Leopold's view, woodsmen could tell that the overabundance of deer was having a detrimental effect on plant life.

Knowing when something is out of kilter is an important ecological skill. The ability to discern when something is starting to get out of kilter is even more helpful. Ranch managers and biologists conduct browse and forb surveys to document how heavily key deer food plants are being browsed as a way to monitor the health of the habitat and prevent potential damage. The combination of overpopulation and excessive browsing is one of the mortal sins of deer management.

The biological effects of overpopulation are well known in agriculture and wildlife management. Farmers know what happens when they plant their crops too densely in the field—all of the plants become stunted due to overcrowding. Foresters know that they must match the number of trees per acre with the soil's productivity. If they allow too many trees to grow, they become spindly, with poor growth rate, and none will ever reach its potential size. Pond managers know they cannot allow too many bass or catfish to populate a body of water, or they outstrip their food supply, leaving a pond full of small, stunted, hungry fish. Ranchers understand that if too many head of livestock are grazed, they will be undernourished, unhealthy, and unproductive and will damage the plants. For these reasons, the proper thinning of crops, timber, fish, livestock, and game is an essential part of responsible management.

Leopold described the ecological damage he observed in the same mountains where he had previously promoted and participated in intensive predator removal. The mule deer population had exploded in response to the eradication of wolves and mountain lions. He wrote, "I have seen every edible bush and seedling browsed, first to anaemic desuetude, and then to death. I have seen every edible tree defoliated to the height of a saddlehorn."[11] Something was out of kilter.

When seedlings are browsed to death, the reproductive cycle ceases. If this happens over a long period of time, those species are essentially exterminated, or nearly so. We are seeing this take place in the Texas Hill Country with Spanish oak, lacey oak, black cherry, wild plum, bigtooth maple, hawthorn, cottonwood, Texas mulberry, white honeysuckle, Carolina buckthorn, and dozens of other palatable browse species. As the old trees and shrubs die of natural causes, there are no young ones to take their place due to over-browsing.

In the 1930s Leopold visited Germany to study forestry and game management practices there. The Germans had noticed a decline in forest productivity and soil health after growing spruce monocultures. With a lack of browse and mast, high populations of roe deer had to be maintained by artificial feeding. The logical solution was to reestablish normal forest diversity. Leopold noted, however: "The German foresters now wish to restore a natural mixture of hardwoods, but the deer won't let them—hardwoods must be fenced to survive the hungry animals."[12] Something was out of kilter and not easily fixed.

A number of biologists, landowners, and ranch managers believe that the overpopulation of deer and exotic browsers, especially axis deer, across vast areas of the Edwards Plateau is one of the most serious ecological problems of the region—worse than feral pigs, invasive grasses, and vast cedar thickets. The damage is not obvious to the casual observer, but the damage is easy to observe for one who knows what the native diversity of shrubs and trees should be. Hillsides that once supported thirty to fifty species of woody plants and possibly one hundred species of forbs now support only one-third or one-half that many. The more desirable species are in decline or already gone; the less desirable species have filled in. Something is out of kilter.

Al Brothers's sage advice for overpopulated white-tail ranges is straightforward: "Kill does until you scare yourself that you have taken too many; then you may be close." This solution is not easy, but the combination of stewardship, woodsmanship, and marksmanship is the place to start.

The Trophy Hunter Who Never Grows Up

The disquieting thing . . . is the trophy-hunter who never grows up, in whom the capacity for isolation, perception, and husbandry is undeveloped, or perhaps lost. He is the motorized ant who swarms the continents before learning to see his own back yard, who consumes but never creates outdoor satisfactions. For him the recreational engineer dilutes the wilderness and artificializes it trophies in the fond belief that he is rendering a public service.[13]

Aldo Leopold, 1938

Aldo Leopold was an avid and enthusiastic hunter by any measure. Most of Leopold's activities for most of his life revolved around hunting and the management of land for game production. Leopold hunted for sport, personal enjoyment, challenge, and instinctive reasons; it was his primary connection to nature. Leopold's father taught him to hunt, and Leopold taught his own children the enjoyment of hunting and sportsmanship. Hunting was at the core of Aldo Leopold. This epigraph above relates Leopold's perspective about an immature form of hunting that he observed and that can still be seen today.

Although Leopold was a lifelong devoted hunter, he was not a trophy hunter, at least not according to our modern perception of the term. The number of birds shot was not his gauge of a successful hunt. The size and score of horns or antlers did not seem to matter to him. Although he certainly had the opportunity to hunt wherever he wished, no photos exist of Leopold posed with trophy-class big game.

However, by his own standards and definitions, Leopold was indeed a first-class trophy hunter. His criteria for trophy hunting were based on genuine hunting skill, sportsmanship, and ethical restraint. Leopold's concept of trophy hunting still resonates with many of today's landowners, hunters, and conservationists. This required the ability to be isolated for long periods; to be at home in the woods, mountains, or marsh; to be able to perceive and appreciate nature and wildness in all of its richness and intricacy; and to have a genuine sense of responsible husbandry and stewardship of animals and habitats. For Leopold and many other like-minded hunters, all hunting can be trophy hunting, regardless of the species hunted and regardless of the kill.

Leopold was a member of the Boone and Crockett Club and never spoke against ethical trophy hunting. However, he was troubled by what he observed in some trophy hunters and some trophy-based wildlife management. And many of today's hunters and landowners will agree.

It is disturbing to see what some aberrant, modern-day trophy hunting has become. For some trophy hunters, it is the quest for the largest, most bizarre, and unnatural antlers possible, regardless of how they were produced. For some, it is merely the pulling of the trigger without the skill, patience, or satisfaction of a challenging hunt. Fortunately, these are anomalies. Most hunters, like Leopold, still hunt in a responsible, mature, and ethical manner and for the same reasons as Leopold—as a vital connection to nature.

The game biologist is partly to blame for the trend toward shallow, artificial hunting. Leopold referred to overly intensive, synthetic forms of wildlife management as "recreational engineering," which he said dilutes the hunting experience. In his book *Game Management*, Leopold said that the value of hunting declines as game management becomes more artificial. Leopold believed that it was not in the public interest to support and enable a greater and greater dependence on artificial game production.

Whether trophy hunting or rabbit hunting, Leopold placed a very high value on the hunter's own personal code of ethics. Leopold wrote, "A peculiar virtue in wildlife ethics is that the hunter ordinarily has no gallery to applaud or disapprove of his conduct. Whatever his acts, they are dictated by his own conscience, rather than by a mob of onlookers. It is difficult to exaggerate the importance of this fact."[14]

Many have marveled at Leopold's uncanny prophetic ability. How could he have known of the ever-increasing trend we see in the motorization of hunting? How could he have been able to predict the degree of artificiality we now see in some forms of game management? Undoubtedly he saw glimpses of it in his day and understood the temptation—and the danger—of making hunting easier.

As we head to the woods, brushlands, and marshes in pursuit of our favorite game, may we keep in mind the real reasons why we hunt. May we savor each moment in the wild and appreciate every aspect of the hunt. If our skills and good fortune are aligned and we are able to take a fine animal, may we be

humble and thankful for the opportunity and blessing. And may we all work
to perpetuate the rich traditions and culture of ethical, responsible hunting.

What Is a Trophy?

[Trophies are] . . .the physical objects that the outdoorsman may seek, find,
capture, and carry away. In this category are wild crops such as game and fish,
and the symbols or tokens of achievement such as heads, hides, photographs, and
specimens. . . .

    The pleasure they give is, or should be, in the seeking as well as the getting.
The trophy, whether it be a bird's egg, a mess of trout, a basket of mushrooms,
the photograph of a bear, the pressed specimen of a wildflower, or a note tucked
into the cairn on a mountain peak, is a certificate. It attests that its owner has
been somewhere and done something—that he has exercised skill, persistence, or
discrimination . . . These connotations which attach to the trophy usually far exceed
its physical value.[15]

Aldo Leopold, 1938

Leopold was an avid hunter, fisherman, naturalist, and outdoorsman by any
measure, but he seemed to care little about our modern concept of the trophy.
He described a far greater array of values to the idea of a "trophy" than what
we normally think today.

    In the fall of the year, many hunters turn their focus toward the pursuit of
a trophy animal. Many definitions are offered of "trophy big game," but nowadays it often boils down to the numerical score of antler or horn dimensions
on male animals. This narrow emphasis is unfortunate and diminishes the
value and perception of a trophy.

    Our views of a trophy have changed in just one generation. Not that many
years ago, a nice chunky six-pound native largemouth was considered a
trophy bass by most fishers. Now it barely gets a nod and is not regarded as
a trophy catch. Likewise, the average Texas deer hunter did not become en-
grossed in calculating antler scores until sometime in the 1980s, and it was
not the primary measuring stick of a trophy.

    Bobwhite authority Dr. Dale Rollins is fond of proclaiming "every quail
is a trophy." That sentiment is certainly valid during times of quail scarcity

but perhaps not so true when quail are plentiful. Rarity does seem to make special things more special.

Not only does rarity elevate the value of a trophy, so also do wildness and the skill involved in the taking. Most will agree that pen-reared quail, pen-raised deer, and put-and-take catfish or trout are inferior to their wild, native counterparts.

In the absence of a physical token, a trophy experience can live in the mind of the hunter, fisher, birdwatcher, botanist, or naturalist. We all have memories of special times outdoors, and we can recall the details and picture them in our minds. My own trophy experiences are numerous, and there is not a trophy room large enough to accommodate them, yet I have no big bucks or large bass hanging on a wall.

My first dove, taken with Dad at my side, as I used his old 12-gauge Stephens single shot, is a trophy memory. Likewise, I have a special memory of Mom taking me out to the field to hunt doves at a time when I required a wheelchair and had my arm in a cast. Shooting doves one-handed is a special challenge, but the love and dedication my mom demonstrated in making the hunt possible is even more special.

The memory of hunting doves with Al Brothers and his three young sons at a dirt tank (stock pond) on Rancho Blanco, the first green-winged teal shot on the Elm Fork of the Trinity, pintails coming into decoys at sunrise, the thrill of hearing thousands of sandhill cranes leaving the roost, and a hundred other memories fill my trophy room.

As Leopold wrote, trophies involve more than pursuing and taking game. A trophy experience can be the finding of a rare grass or the discovery of heart-leaf hibiscus in full bloom in a thicket of guajillo. These "minor trophies" can be just as memorable and special as rattling up a big South Texas whitetail. Collecting deer hair, squirrel tail, and turkey feathers, tying them onto a number 4 hook in the form of a muddler, and sending the perfect fly cast over a sunken log on the Llano, enticing the big strike of a Guadalupe bass—these are all pieces of a trophy experience.

Another trophy experience was watching my young daughter delight in catching sunfish by the dozen on a cane pole and then watching her children do the same twenty-five years later. And there is the memory of my then-eight-year-old-son winning the annual family carp tournament and barely being able to hold up the huge fish he caught.

A trophy experience involves a lot more than pulling the trigger. The skills required, the excitement, and anticipation are all facets of a genuine trophy hunt, regardless of the kill.

The memories of fishing with my brother and Grandpa Stephan on the Trinity River are still fresh and clear. Even when we did not catch a big tub full of catfish, the memories of being with Grandpa and watching the river are still trophy-quality experiences. Seeing the monstrous alligator gar rolling in the turbid waters is etched in my mind as if it was yesterday, although it happened over a half century ago.

You can be a trophy hunter with or without a shotgun, bow, fly rod, camera, or binoculars. With natural curiosity, eyesight, or hearing, anyone can have a trophy experience with nature. A trophy experience can happen while hunting, fishing, birdwatching, camping, hiking, canoeing, or sitting alone on a ridge watching a distant thunderstorm, sunrise, or sunset.

For many outdoorsmen and outdoorswomen, a certain amount of grace is involved in the collection of trophies. We recognize that trophy experiences are mostly undeserved and unexpected moments that occur when we happen to be in the right place at the right time to observe or find something special, rare, or unique. As we enter the next hunting season, let us not forget that the true measure of a trophy lies in the challenge of seeking the game and the experiences we collect along the way.

May every day spent in the woods or beside the river be a trophy day, whether or not we pull the trigger or set the hook. May your hunting season never end and your trophies be many and special.

Artificiality

The recreational value of a head of game is inverse to the artificiality of its origin; and hence to the intensiveness of the system of game management which produced it.[16]

Aldo Leopold, 1931

Being called artificial is one of the most degrading things that can be said about someone. To be artificial is to be shallow, fake, insincere, or superficial—qualities admired by no one.

Artificial leather is inferior to real leather, and polyester fabric is nothing like cloth made of natural fiber. Imitation wood does not have nearly the same quality or beauty as real wood. The mechanical bull at Gilley's is not the same as trying to stay on a big bad Brahman at the rodeo. The list could go on—artificial flowers, artificial grass, nondairy ice cream, turkey bacon, fake diamonds, fake fur, fake meat, imitation maple syrup, cultured marble, and more. In today's world of synthetics, we are inundated with artificial substitutes that are not like the real thing.

In wildlife management, a trend toward artificiality was prevalent in the days of Leopold and is still with us today. The proponents thought that anything to get more game was justified regardless of the means of producing it. The big push at that time was artificial propagation of quail, ducks, and pheasants to make up for the lack of natural wild populations. Leopold did not hide his indignation at the practice. He preferred to manage natural habitat so that wild game would thrive in a wild setting to support a genuine hunting experience.

Leopold spoke of two divergent groups of wildlife managers. He said Group A is content to produce wildlife with the same level of intensity the farmer uses to produce cotton, corn, poultry, or pigs. Group B also desires wildlife productivity but prefers to manage in a natural setting by tweaking the habitat rather than resorting to artificial means. These two groups are still active today.

Most serious quail hunters scorn the thought of pursuing pen-raised quail or hunting along baited roads. Likewise, hunting exotic or native big game in a large free-range setting is a lot different than shooting the same animal in a small enclosure. Catching wild channel catfish from a flowing river is a lot different than catching the same species at a catfish farm. A serious birder will not receive the same satisfaction from seeing birds at a backyard feeder as from observing them in the wild. Encountering a mountain lion or rattlesnake in the wild is a completely different experience than seeing them at the zoo.

All degrees of artificiality exist in wildlife management, and some artificiality is inevitable. All landowners and wildlife enthusiasts must decide how much they are willing to tolerate and where to draw the line. For the new hunter, fisher or birdwatcher, a higher degree of artificiality may be warranted as a way to get started in the sport and generate interest. As skills develop, most will prefer a more natural experience over the artificial. The recreational value of hunting or fishing is a lot more than pulling the trigger or landing the fish. Recreational quality involves enjoying the entire process of planning, practicing, honing skills, training dogs, dreaming, anticipating, waiting, watching, listening, thinking, stalking, calling, casting, and rattling, as well as the excitement and nervousness and then the satisfaction and reliving and retelling of the experience for decades. Remembering the big one that got away can be just as memorable as a successful ending.

Leopold claimed that artificiality diminishes the recreational value of hunting. Many will agree, and some will disagree, and opinions are strong on both sides. Charlie Granstaff, a Menard County, Texas, landowner, hunter, and conservationist, speaks for many when he says that "artificial propagation of deer is like steroids in baseball—it's not the real thing and hurts the sport."

Yet others argue that artificial deer farming is within the scope of landowner rights and that it meets a demand for larger antlers that some hunters are looking for.

Many will admit that something special is diminished for every new degree of artificiality we introduce into hunting, fishing, and other forms of outdoor recreation. Just because we have the technological ability to enhance and multiply what nature can do does not mean it is always a good idea. The degree of intensity and artificiality warranted in the raising of agricultural commodities does not always translate nicely into the world of wildlife management.

Wildlife Management and Agriculture

Game management is the art of making land produce sustained annual crops of wild game for recreational use. . . . Effective [wildlife] conservation requires . . . a deliberate and purposeful manipulation of the environment . . . [and] this manipulation can only be carried out by the landowner, and that the private landowner must be given some kind of an incentive for undertaking it. . . .

Game management is a form of agriculture.[17]

Aldo Leopold, 1933

Aldo Leopold's book *Game Management*, published in 1933, was the first text on the science of managing wildlife. The book was not meant to be inspirational or philosophical as some of his other writings were, but it laid a solid framework for the new discipline. In this book, Leopold makes a clear case that agriculture and wildlife management are close allies.

People today sometimes make the mistake of thinking that wildlife conservation is entirely different from and even contrary to agriculture. From time to time we hear comments from both the agricultural side and the wildlife side denigrating the other. Some agriculturalists look down on those who are primarily interested in wildlife, and some wildlife-oriented people look down upon the agricultural community as being anti-wildlife. This is unfortunate because both sides share much common ground. In fact, both sides need each other.

Painted buntings make use of water that was developed primarily for livestock. Many ranch management practices benefit wildlife as much as they do livestock, demonstrating that wildlife management and agriculture are close allies.

Although there are some basic differences between raising crops and livestock and managing wildlife, the similarities are more profound than the differences. In a state like Texas with mostly private land, wildlife and agriculture are inseparable, compatible, cooperative, and even synergistic in most cases. Each can thrive in association with the other.

Aldo Leopold is most well known as a wildlife manager, but he was also an agriculturalist and a landowner. His formal schooling and much of his career was in forestry—the growing and harvesting of trees for commercial timber products. Leopold realized that, in many places, private farms, ranches, and forests are the primary abode of wildlife and that agricultural people are the primary wildlife managers. Of all the things that Leopold wrote about, it is clear that he cherished his own farm as much or more than anything else. Leopold's farm was not large enough to make a living on, but he poured himself into improving, managing, and restoring the poor Wisconsin sandy-land farm.

Leopold promoted active and deliberate manipulation of the environment in ways that are similar to what agriculture must do. Leopold did not advocate a hands-off, preservationist approach to wildlife management. He believed in using the land and deriving products, profit, and benefit from it. Leopold considered wildlife as both a by-product of other agricultural uses as well as a primary product of the land where that was the landowner's intent.

Leopold's most famous thesis about game management is proof that agriculture and wildlife management are similar. His oft-quoted thesis takes the tools and techniques commonly used in agriculture at the time (axe, plow, cow, fire, and gun) and proposed that they could be used creatively to produce sustained crops of wildlife. The validity of this thesis is affirmed by the fact that it is still widely used by today's landowners for the purpose of conservation, habitat management, forest management, and livestock ranching.

Leopold was sometimes hard on abuses he observed in farming, ranching, and forestry, but he was just as hard on abuses he saw in wildlife management and hunting. Leopold was quick to criticize the farmer who allowed his land to erode, and he was quick to criticize the rancher for overgrazing ranges. He was equally quick to criticize the unscrupulous sportsman, the immature trophy hunter and the game manager who allowed damage to the habitat by deer overpopulation.

There can be a blurry line between agriculture and wildlife management. Both require some manipulation, but, in Leopold's eye, wildlife management minimizes artificial manipulation to keep the essential character of wildness intact. Most wildlife managers in Texas seem to concur with Leopold's view.

If Leopold were still alive, we can be certain that he would be advocating that farmers, ranchers, wildlife managers, and hunters continue to find common ground and work together. To the farmer or rancher, he would say "Get to know your new neighbor who is interested in hunting." To the new landowners who bought land for wildlife and recreation, Leopold would advise a cooperative and friendly relationship with their agricultural neighbors. Each can learn from the other, and each can assist the other.

Leopold was a strong advocate of economic incentives and market-based compensation, long before these became the norm. Without economic returns, he knew wildlife management would never thrive. Leopold would find that system thriving in Texas. Because of the financial incentives involved in hunting and other forms of outdoor recreation, Texas wildlife is in good shape—cared for and managed by agricultural people and private land stewards.

Wildlife is an agricultural product of the land, just as cotton and cattle are. While they are not exactly the same, they are similar. The sustained production of crops, livestock, timber, and wildlife requires manipulation of the environment, even though the intensity of the manipulation varies. All these forms of agriculture share a common need for conservation, sustainability, and stewardship.

NOTES

In the notes below, quotes by Leopold give the original source, usually followed by an abbreviation in brackets indicating where the material has been reprinted and is more accessible.

Abbreviations

[ALO]   Julianne Lutz Newton, *Aldo Leopold's Odyssey* (Washington, DC: Island Press, 2006).

[ALS]   David E. Brown and Neil B. Carmony, *Aldo Leopold's Southwest* (Albuquerque: University of New Mexico Press, 1995).

[ASCA] Aldo Leopold, *A Sand County Almanac and Sketches Here and There* (New York: Oxford University Press, 1949).

[ASCAO] Aldo Leopold, *A Sand County Almanac and Other Writings on Ecology and Conservation,* ed. Curt Meine (New York: Library of America, 2013).

[CASCA] Aldo Leopold, *Companion to "A Sand County Almanac,"* ed. J. Baird Callicott (Madison: University of Wisconsin Press, 1987).

[EAL]   Aldo Leopold, *The Essential Aldo Leopold*, ed. Curt Meine and Richard Knight (Madison: University of Wisconsin Press, 1999).

[FHL]   Aldo Leopold, *For the Health of the Land*, ed. J. Baird Callicott and Eric T. Freyfogle, (Washington, DC: Island Press, 1999).

[GM]   Aldo Leopold, *Game Management* (New York: Charles Scribner's Sons, 1933).

[RMG]   Aldo Leopold, *The River of the Mother of God and Other Essays by Aldo Leopold*, ed. Susan L. Flader and J. Baird Callicott (Madison: University of Wisconsin Press, 1991).

[RR]   Aldo Leopold, *Round River: From the Journals of Aldo Leopold*, ed. Luna B. Leopold (New York: Oxford University Press, 1953).

Chapter 1

1. Quoted in C. D. Meine, *Aldo Leopold: His Life and Work* (Madison: University of Wisconsin Press, 1988), 81–82.

2. Aldo Leopold letter to Robert Kleberg, February 10, 1947, Aldo Leopold Papers, Series 9/24/10-3, Box 10, Folder 3, University of Wisconsin–Madison.

3. Aldo Leopold letter to Val Lehmann, March 12, 1947, Aldo Leopold Papers, Series 9/24/10-3, Box 10, Folder 3, University of Wisconsin–Madison Archives.

4. Aldo Leopold Letter to Val Lehmann, February 14, 1947, Aldo Leopold Papers, Series 9/24/10-3, Box 10, Folder 3, University of Wisconsin–Madison Archives.

Chapter 2

1. Leopold, "A Biotic View of Land," *Journal of Forestry* 37, no. 9 (September 1939). [RMG, 267].

2. Leopold, "Farmer-Sportsman: A Partnership for Wildlife Restoration," *Transaction of the Fourth North American Wildlife Conference*, February 1939. [EAL, 68]

3. Leopold, "Conservation," from "A Survey of Conservation" manuscript, 1938. [RR, 146–47]

4. Leopold, "The Farmer as a Conservationist," *American Forests* 45, no. 6 (June 1939). [RMG, 261]

5. Ralph Waldo Emerson, *Fortune of the Republic* (Boston: Houghton, Osgood, 1879), 8.

6. Leopold, "Lakes in Relation to Terrestrial Life Patterns," in James G. Needham et al., *A Symposium on Hydrobiology* (Madison: University of Wisconsin Press, 1941). [ASCAO, 447]

7. Leopold, "A Biotic View of Land," *Journal of Forestry* 37, no. 9 (September 1939). [RMG, 268]

8. Leopold, "Conservation," from "A Survey of Conservation" manuscript, 1938. [RR, 147–48]

9. Leopold, "Cheat Takes Over," *The Land*, 1, no. 4 (Autumn 1941). [ASCA, 167]

10. Leopold, "Conservation: In Whole or in Part?," manuscript, November 1, 1944. [RMG, 310]

11. Leopold, "Wilderness as a Land Laboratory," *Living Wilderness* 6 (July 1941). [ASCA, 274]

12. Leopold, "Song of the Gavilan," *Journal of Wildlife Management* 4, no. 3 (July 1940). [ASCA, 162]

Chapter 3

1. Leopold, *A Sand County Almanac and Sketches Here and There* (New York: Oxford University Press, 1949), 223.

2. J. L. Merrill, "The XXX Ranch: Managing Range for Ecology and Economy," in *Using Our Natural Resources: 1983 Yearbook of Agriculture* (Washington, DC: US Department of Agriculture, 1983), 86.

3. Leopold, *A Sand County Almanac*, 202.

4. Leopold, "Conservation," typewritten draft, filed with correspondence dated August 8, 1946. [EAL, 166]

5. Leopold, "Conservation," from "A Survey of Conservation" manuscript, 1938. [ASCA, 202]

6. Leopold, "Conservationist in Mexico," *American Forests* 43, no. 3 (March 1937). [ALS, 207].

7. Leopold, "The Ecological Conscience," *Bulletin of the Garden Club of America*, September 1947. [ASCA, 224–25]

8. Leopold, "The Farmer as a Conservationist," *American Forests* 45, no. 6 (June 1939). [RMG, 264]

9. Leopold, *A Sand County Almanac*, 221.

10. Leopold, "Notes on Proposed Centennial Symposium on Ecological Conservation," undated, unpublished draft, 1947. [EAL, 166]

Chapter 4

1. Leopold, "A Man's Leisure Time," manuscript, October 15, 1920. [EAL, 257]

2. Leopold, "The Role of Wildlife in Liberal Education," paper presented at the seventh North American Wildlife Conference, April 8–10, 1942. [RMG, 302]

3. Leopold, "Home Range," *Wisconsin Conservation Bulletin* 8, no. 9 (September 1943). [ASCA, 81]

4. John E. Weaver, *North American Prairie* (Lincoln, Neb.: Johnson, 1954), 325.

5. Leopold, "The Farmer as a Conservationist," *American Forests* 45, no. 6 (June 1939). [RMG, 262–63]

6. Leopold, *A Sand County Almanac and Sketches Here and There* (New York: Oxford University Press, 1949), 48–50.

7. Leopold, "The Farmer as a Conservationist." [RMG, 257]

8. Leopold, "Academic and Professional Training in Wildlife Work," *Journal of Wildlife Management* 3, no. 2 (April 1939). [EAL, 68–69]

9. Leopold, "Teaching Wildlife Conservation in Public Schools," *Transactions of the Wisconsin Academy of Sciences, Arts and Letters* 30 (1937). [EAL, 260]

10. Leopold, "A Plea for Recognition of Artificial Works of Forest Erosion Control Policy," *Journal of Forestry* 19, no. 3 (March 1921). [ALO, 65]

11. Leopold, "Sky Dance of Spring," *Wildlife Conservation on the Farm*, September 1941. [ASCA, 32–33]

Chapter 5

1. Leopold, "The Farmer as a Conservationist," *American Forests* 45, no. 6 (June 1939). [RMG, 257]

2. Leopold, "Conservation," from "A Survey of Conservation," manuscript, 1938. [RR] 145–47]

3. Leopold, "Conservation: In Whole or in Part?," manuscript, November 1, 1944. [RMG, 318–19]

4. Leopold, "The Farmer as a Conservationist." [RMG, 258]

5. Leopold, "Conservation: In Whole or in Part?" [RMG, 316]

6. Leopold, "Land Use Democracy," *Audubon*, 44, no. 5 (September–October, 1942). [RMG, 298]

7. Leopold, "Conservation," from "A Survey of Conservation," manuscript, 1938. [RR, 152–53]

8. Leopold, "The Ecological Conscience," *Bulletin of the Garden Club of America*, September 1947. [RMG, 344]

9. Leopold, foreword to *Great Possessions* (unpublished), Leopold Papers 6B16, University of Wisconsin–Madison Archives. [CASCA, 282]

10. Leopold, *The University and Conservation of Wisconsin Wildlife*, University of Wisconsin, ser. no 2211, Science Inquiry Publication no 3, February 1937. [EAL, 259]

11. Leopold, *A Sand County Almanac and Sketches Here and There* (New York: Oxford University Press, 1949), 221.

12. Leopold, *A Sand County Almanac*, 225.

13. Leopold, Conservation Economics," *Journal of Forestry* 32, no. 5 (May 1934). [RMG, 202]

Chapter 6

1. Leopold, "Grand-Opera Game," manuscript, summer 1932. [RMG, 172]

2. Leopold, "The Farmer as a Conservationist," *American Forests* 45, no. 6 (June 1939). [RMG, 263]

3. Leopold, "Conservation Economics," *Journal of Forestry* 32, no. 5 (May 1934). [RMG, 197]

4. Leopold, *A Sand County Almanac and Sketches Here and There* (New York: Oxford University Press, 1949), 68.

5. Leopold, "The Virgin Southwest," manuscript, May 6, 1933. [RMG, 178]

6. Leopold, "Pioneers and Gullies," *Sunset* 52, no. 5 (May 1924). [RMG, 110]

7. Leopold, "A Plea for Recognition of Artificial Works of Forest Erosion Control Policy," *Journal of Forestry* 19, no. 3 (March 1921). [ALO, 65]

8. Leopold, *A Sand County Almanac*, 221–22.

9. Leopold, *A Sand County Almanac*, 70–72.

10. Leopold, "Some Thoughts on Recreational Planning," *Parks and Recreation* 18, no. 4 (December 1934). [EAL, 160]

11. Leopold "Conservation Economics," *Journal of Forestry* 32, no. 5 (May 1934). [RMG, 202]

12. Leopold, "Coon Valley: An Adventure in Cooperative Conservation," *American Forests* 41, no. 5 (May 1935). [RMG, 220]

Chapter 7

1. Leopold, "Planning for Wildlife," manuscript, September 26, 1941. [EAL, 95]

2. Leopold, review of Stanly P. Young and Edward H. Goldman, *The Wolves of North America* (Washington, D.C.; American Wildlife Institute, 1944), *Journal of Forestry* 43, no. 1 (January 1945). [ALS, 226]

3. Leopold, "Wild Lifers vs. Game Farmers: A Plea for Democracy in Sport," *Bulletin of American Game Preservation Association* 8, no. 2 (April 1919). [ALS, 57]

4. Leopold, "Wanted—National Forest Game Refuges," *Bulletin of the American Game Protective Association* 9, no. 1 (January 1920). [EAL, 52]

5. Leopold, "The Game Situation in the Southwest," *Bulletin of the American Game Protective Association* 9, no. 2 (April 1920). [EAL, 53]

6. Leopold, "Game Management in the National Forests," *American Forests* 36, no. 7 (July 1930). [ALS, 128]

7. Leopold, "The Farmer as a Conservationist," *American Forests* 45, no. 6 (June 1939). [RMG, 257]

8. Leopold, "Farmer-Sportsman: A Partnership for Wildlife Restoration," *Transaction of the Fourth North American Wildlife Conference*, February 1939). [EAL 68]

9. Leopold, *A Sand County Almanac and Sketches Here and There* (New York: Oxford University Press, 1949), 130.

10. Leopold, "The Farmer as a Conservationist." [RMG, 257]

11. Leopold, *A Sand County Almanac*, 130.

12. Leopold, "Farm Game Management in Silesia," *American Wildlife* 25, no. 5 (September–October 1936). [FHL 56]

13. Leopold, "Conservation Esthetic," *Bird Lore* 40, no. 2 (March–April 1938). [ASCA, 176]

14. Leopold, *A Sand County Almanac*, 178.

15. Leopold, "Conservation Esthetic." [ASCA, 168–69]

16. Leopold, "Game Methods: The American Way," *American Game* 20, no. 2 (March–April 1931). [RMG, 158]

17. Leopold, *Game Management* (New York: Charles Scribner's Sons, 1933), 3, 21, 395.

INDEX

Page numbers in *italic* refer to illustrations.